Globular clusters are roughly spherical, densely packed groups of stars found around galaxies. Most globular clusters probably formed at the same time as their host galaxies. They therefore provide a unique fossil record of the conditions during the formation and early evolution of galaxies.

This volume presents the first comprehensive review of globular cluster systems. It summarises their observed properties, and shows how these constrain models of the structure of stars, the formation and evolution of galaxies and globular clusters, and the age of the Universe.

For graduate students and researchers, this timely volume provides the definitive reference on globular cluster systems.

GLOBULAR CLUSTER SYSTEMS

Cambridge astrophysics series

Series editors

Andrew King, Douglas Lin, Stephen Maran, Jim Pringle and Martin Ward

GLOBULAR CLUSTER SYSTEMS

KEITH M. ASHMAN

Department of Physics and Astronomy, University of Kansas

STEPHEN E. ZEPF

Department of Astronomy, University of California, Berkeley

CAMBRIDGE
UNIVERSITY PRESS

CAMBRIDGE UNIVERSITY PRESS
Cambridge, New York, Melbourne, Madrid, Cape Town, Singapore, São Paulo, Delhi

Cambridge University Press
The Edinburgh Building, Cambridge CB2 8RU, UK

Published in the United States of America by Cambridge University Press, New York

www.cambridge.org
Information on this title: www.cambridge.org/9780521550574

© Cambridge University Press 1998

First published 1998
This digitally printed version 2008

A catalogue record for this publication is available from the British Library

Library of Congress Cataloguing in Publication data

Ashman, Keith M., 1963–
Globular cluster systems / Keith M. Ashman, Stephen E. Zepf.
          p.   cm.
Includes bibliographical references and index.
ISBN 0 521 55057 2
1. Stars – Globular clusters.   2. Galaxies – Formation.
3. Astrophysics.   I. Zepf, Stephen E., 1963–   II. Title.
QB853.5.A84   1998
523.8′55–dc21   97-17391   CIP

ISBN 978-0-521-55057-4 hardback
ISBN 978-0-521-08783-4 paperback

Contents

Preface

This book gives a comprehensive account of the current understanding of globular clusters and globular cluster systems. We describe the observed properties of these objects and discuss the theoretical ideas that are believed to account for their properties. A key theme of the book is to link the study of globular cluster systems to other areas of astrophysics, such as galaxy formation and evolution. The text is designed for graduate students working in globular cluster research, and as a resource for research astronomers active in this area. We also hope that the connections we emphasize between globular cluster systems and other astrophysical systems will make the book useful to researchers in related fields.

Many people have been instrumental in bringing this book about. The original suggestion that we embark on the project came from Simon Mitton, who accurately described the potential pitfalls and rewards of writing a book. Our editor, Adam Black, has patiently provided advice and encouragement. We are grateful to our copy-editor, Sheila Shepherd, for her careful reading of the original manuscript and her extremely helpful suggestions and corrections, and many others at Cambridge University Press who assisted in bringing this book to fruition. Our own work in the field would not have been possible without our collaborators: Christina Bird, Terry Bridges, David Carter, Alberto Conti, Jayanne English, Kenneth Freeman, Doug Geisler, David Hanes, Ray Sharples and Joseph Silk. Many other colleagues have provided considerable help in preparing the book, suggesting revisions to earlier versions and providing us with data and figures. In particular, we would like to thank Barbara Anthony-Twarog, Michael Bolte, Roberto Buonanno, Adrienne Cool, George Djorgovski, Duncan Forbes, Doug Geisler, Bill Harris, Ivan King, John Laird, Peter Leonard, Georges Meylan, Ata Sarajedini, Graeme Smith, Peter Stetson, Jeremy Tinker, Scott Trager, Scott Tremaine, Virginia Trimble, Bruce Twarog, Sidney van den Bergh, Brad Whitmore, and Robert Zinn. Sidney van den Bergh deserves special mention for carefully reading and commenting on the entire manuscript. The astronomers listed above, and many others, have also provided us with a constant supply of preprints and other information that helped us to keep abreast of the most recent developments in the field.

Much of our work in the areas described in the book has been supported by fellowships and grants from NASA through the Space Telescope Science Institute. We have also benefitted from a NATO collaborative grant and made significant progress on the manuscript at the Aspen Center for Physics in June, 1996. The stimulating working environment at Berkeley also helped the completion of the book during collaborative visits. Most importantly, this work would not have been completed without the support and encouragement of our wives, Christina Bird and Christina Schwarz.

1

Introduction

'Perhaps the most wonderful of all the star clusters are those in which hundreds upon hundreds of faint stars are all gathered together in the shape of a globe.' Although this description of globular clusters by Reverend James Baikie (1911) is numerically faulty, it summarizes the visually striking features of these objects accurately. Globular clusters are characterized by high central stellar densities, and tend to be extremely round. Figure 1.1 shows an image of the Milky Way globular cluster M92 illustrating these features. Globular clusters like M92 formed early in the history of the universe and contain some of the oldest stars known. Systems of globular clusters appear to surround all bright galaxies, as well as many dwarf galaxies.

Fig. 1.1. The globular cluster M92 (Lick Observatory).

Individual globular clusters and globular cluster systems are of interest in their own right, but they also provide unique insights into a wide range of astrophysical processes and systems. As individual objects, they constitute isolated laboratories in which many aspects of stellar evolution and dynamics can be studied. When considered as a system, the globular clusters surrounding a galaxy provide a fossil record of the dynamical and chemical conditions when the galaxy was in the process of formation. Thus observations of globular cluster systems can be used to constrain models of the formation and evolution of galaxies. Perhaps the most renowned impact of globular cluster research on other fields of astronomy is provided by age estimates of Milky Way globulars, which provide a minimum age for the universe.

The interplay between globular cluster research and other areas in astronomy is not a recent development. One historical example is provided by studies of the Milky Way globular cluster system and the role they played in uncovering the size and shape of the Galaxy. Early in the twentieth century, Kapteyn investigated the size and structure of the Milky Way (then regarded as the entire universe) by measuring the positions and magnitudes of stars on photographic plates. He concluded that stars in the Milky Way had a somewhat flattened distribution with the Solar System close to the center.

A key clue that eventually produced a major revision in this model was contained in the observations of globular clusters carried out by John Herschel in the 1830s. He noticed that a large number of these clusters occurred in a relatively small portion of the sky in the direction of Sagittarius. In 1909, Karl Bohlin remarked on the related fact that the majority of globular clusters are found in one half of the sky. He suggested that the Milky Way globular cluster system was distributed symmetrically about the Galactic center, so that the observed distribution implied that the Solar System was displaced from the center of the Milky Way. There were several oddities in Bohlin's model which obscured this important insight of moving the Solar System away from the Galactic center.

The crucial advance in this area was provided in 1918 by Harlow Shapley, who observed variable stars in globular clusters which he assumed were Cepheids. His calibration of the absolute magnitude of these variables allowed Shapley to derive distances to globular clusters. (It was later realized that the RR Lyrae variables in globular clusters are fainter than Cepheids, and that Shapley had somewhat overestimated globular cluster distances as a result.) Like Bohlin, Shapley assumed that the spatial distribution of globular clusters was symmetric about the Galactic center. With this assumption, Shapley's distances enabled him to estimate the overall size of the system, as well as the displacement of the Solar System from the Galactic center. Not only did Shapley's results overturn the accepted orthodoxy of the Solar System located towards the inner regions of the Milky Way, they also gave a size of the Milky Way an order of magnitude greater than the Kapteyn model.

We now know that, to within a factor of two or so in scale, Shapley's model was basically correct. Kapteyn and Herschel had not fully accounted for interstellar extinction produced by dust in the Galactic plane, which gave the impression that the number of stars in the plane of the Galaxy fell off fairly uniformly in all directions. The large number of globular clusters in Sagittarius observed by Herschel marks the direction of the Galactic center.

Globular clusters have also played an important role in our understanding of the differences between stellar populations and the resulting implications for galactic structure and evolution. In the 1940s, Baade pioneered work on stellar populations, noting the distinction between Population I stars, such as those in the solar neighborhood, and Population II stars that he identified in the bulge of M31 and its satellite galaxies M32 and NGC 205. Importantly, Baade noted that stars in Galactic globular clusters belonged to Population II. Subsequent observations by Arp, Baum and Sandage led to the conclusion that the stars in Milky Way globular clusters, and thus Population II stars in general, were extremely old. The great age of Milky Way globular clusters is one of the attributes that makes them valuable as probes of the formation of the Galaxy, since they probably formed when the Milky Way itself was forming. Because all their constituent stars are at the same distance, globular clusters are also invaluable as testbeds for understanding stellar structure and evolution, and for serving as the basis for models of stellar populations.

The observational work on stellar populations and the study of globular cluster color–magnitude diagrams in particular led to spectacular advances in stellar evolution theory. In the 1940s and 1950s, laboratory work in the field of nuclear physics allowed calculations to be made of energy generation and transport in stars. Much of the influential work in this area was carried out by Schwarzschild, Hoyle, Henyey and their collaborators. The great age of Milky Way globular clusters allowed a quantitative comparison between the observed colors and luminosities of evolved stars with theoretical predictions. One of the key developments was the realization that nuclear reactions in stars brought about chemical changes. It was found that evolution off the main sequence was a consequence of hydrogen being exhausted in the central stellar regions. Massive stars evolve more rapidly because they use up their hydrogen on a shorter timescale. These and other ideas, which were tried and tested through comparison with observations of globular cluster stars, provide the basis for modern stellar evolution theory.

Another valuable property of globular clusters is their ubiquity. Shortly after the realization that spiral 'nebulae' were galaxies similar to the Milky Way, Hubble detected globular clusters in M31. It was not until the 1970s, however, that globular clusters were observed in any significant number around galaxies beyond the Local Group. Since that time, it has become apparent that all bright galaxies probably have globular cluster systems, as do many dwarf galaxies. The presence of relatively accessible tracers of the early conditions in galaxies provides a key tool in studying the galaxy formation process.

One important development in recent years is the evidence that dense, massive star clusters are currently forming in certain environments. It has been suggested that these objects are young globular clusters. This idea is not universally accepted, both because current observations are not definitive and possibly because the notion of *young* globular clusters flies in the face of the traditional view of globular clusters as ancient objects. However, if globular clusters are forming at the present epoch, we will have the opportunity to study the formation process directly. It seems inevitable that this will greatly enhance our understanding of how and why globular clusters form, as well as deepening our knowledge of the galaxy formation process to which globular cluster formation is intimately related.

In this book, our primary focus is a discussion of globular cluster *systems*. We describe the observational properties of such systems and the theoretical inferences and constraints that can be obtained from these observations. However, to put matters in context, we first describe the internal properties of globular clusters. This material is covered in Chapter 2, where we concentrate on the characteristic properties of globular clusters in the Milky Way. In Chapter 3 we look at the Milky Way globular clusters as a system, and discuss how the properties of this system have influenced (and continue to influence) ideas on the formation of our Galaxy. The globular cluster systems of galaxies in the Local Group and slightly beyond are discussed in Chapter 4, whereas more distant extragalactic globular cluster systems are addressed in Chapter 5. Correlations between the properties of globular cluster systems and properties of their host galaxies are also discussed in Chapter 5. Chapter 6 provides an overview of theories of galaxy formation and discusses how such theories are constrained by the observed properties of globular cluster systems. Formation models of globular clusters and globular cluster systems are covered in Chapter 7. A summary and some speculations concerning the future directions of globular cluster research are presented in Chapter 8.

Throughout this book, we attempt to highlight areas where globular cluster research has had an important impact on other astronomical fields, and, perhaps more importantly, where the interplay continues to advance astronomical understanding.

2

Properties of globular clusters

Like most astronomical objects, globular clusters exhibit a range of properties and characteristics, but certain features are common to the majority of them. Due to their relative proximity, globular clusters within the Milky Way are the best-studied, and most of the 'typical' properties described in this chapter are based on observations of these objects. Unless otherwise stated, the data used in this and the following chapter are taken from the McMaster catalog described by Harris (1996; see also Harris and Harris 1997). Differences between the globular clusters of the Milky Way and other galaxies are summarized at the end of this chapter.

2.1 Color–magnitude diagrams

The luminosity and temperature of a star are dependent on its mass, age, and chemical composition. Color–magnitude diagrams of globular clusters have long been the subject of intensive study because they reflect these fundamental properties of the constituent stars. Figure 2.1 shows the color–magnitude diagram of M5 and illustrates a number of the basic features of globular cluster color–magnitude diagrams. These include the main sequence, the giant branch, and the horizontal branch, each of which is discussed in the following subsections.

2.1.1 Main sequence

One of the key features of globular clusters is the well-defined main sequence extending from the turn-off to fainter magnitudes and redder colors (see Figure 2.1). Globular cluster stars on the main sequence derive their energy from the conversion of hydrogen to helium in the stellar core. The low-luminosity end of the main sequence shown in Figure 2.1 is determined by the magnitude limit of the observations. (There is also a theoretical lower limit to the main sequence corresponding to a stellar mass around 0.08 $M_\odot$, below which hydrogen burning no longer takes place in the stellar core.) A characteristic feature of the color–magnitude diagrams of Galactic globular clusters is that the turn-off of the main sequence occurs at fainter luminosities than for most star clusters in the solar neighborhood. This was first established in the early 1950s by Sandage, Arp, and others (e.g. Arp *et al.* 1953; Sandage 1953). It was soon realized that the fainter luminosity of the main-sequence turn-off indicated that these globular clusters are old (Sandage and Schwarzschild 1952; Hoyle and Schwarzschild 1955). In the cores of more massive stars, hydrogen is exhausted more rapidly,

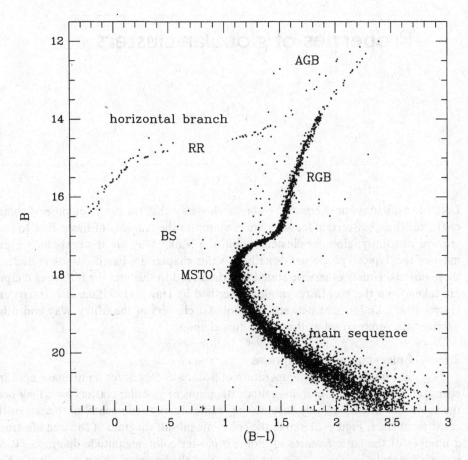

Fig. 2.1. The color–magnitude diagram of M5. The horizontal branch and main sequence are labeled. Also shown are: the RR Lyrae gap or instability strip (RR); the Red Giant Branch (RGB); the asymptotic giant branch (AGB); the main-sequence turn-off (MSTO); and blue stragglers (BS). (From data supplied by M. Bolte.)

so that older stellar populations have main-sequence turn-offs at lower stellar masses and thus luminosities. As discussed in more detail in Section 2.2, studies of the location of the main-sequence turn-off in well-studied Galactic globular clusters give turn-off masses of about 0.8 M$_\odot$ and corresponding ages of roughly 15 Gyr.

In addition to age, the color and temperature of stars along the main sequence are determined by a number of parameters, the most important being chemical composition. Higher metallicity clusters have fainter and redder main-sequence turn-offs at a fixed age compared to lower metallicity clusters. As first discussed by Reiz (1954), Schwarzschild *et al.* (1955), and others, the dominant physical process responsible for this dependence on metallicity is the increased opacity in the stellar atmospheres due to electrons from metals. This allows stars to maintain equilibrium at a lower temperature. Line blanketing in atmospheres also contributes to producing redder colors for higher metallicity stars. More recent stellar evolution studies provide values of the mass, luminosity and color of stars at the main-sequence turn-off as a function of age and metallicity (e.g. Straniero and Chieffi 1991).

The sharp main-sequence turn-off of M5 in Figure 2.1 is typical of globular cluster color–magnitude diagrams, indicating that the stars within an individual globular cluster all formed at roughly the same time. Another well-studied cluster is M92, which we discuss in detail in Section 2.2 below. The form of the main-sequence turn-off in M92 limits the age spread between the constituent stars to about 2.4% of the age of this cluster, or around 0.4 Gyr (Stetson 1993).

Figure 2.1 illustrates another characteristic of globular cluster main sequences – they are very narrow. This narrowness indicates that all the stars in the globular cluster have a very similar chemical composition. Constraints on chemical inhomogeneities are discussed in more detail in Section 2.3. The narrowness of the main sequence also constrains the fraction of binary stars within globular clusters. Unresolved binaries are expected to produce a population just above the main sequence, since the combined luminosity of the two stars exceeds that of a single star at the same color. (However, if the primary and secondary have significantly different masses, this effect is difficult to detect.) There are globular clusters where such a population of binaries has been detected, such as NGC 288, where the inferred binary fraction is around 10% (Bolte 1992). Future observations, particularly with the *Hubble Space Telescope* (hereafter *HST*), are likely to result in better constraints on the binary fraction in more globular clusters (see Rubinstein and Bailyn 1997).

2.1.2 Red giant branch

The red giant branch (RGB) in globular clusters extends from the subgiants which connect it to the main sequence to brighter magnitudes and redder colors until the tip of the RGB is reached (see Figure 2.1). Observational properties and characteristics of the RGB, as well as a survey of earlier literature, are given by Stetson (1993). While the detailed evolution of stars on the RGB is a complicated topic, the salient feature is that such stars possess hydrogen-burning shells, which advance outwards as they ascend the giant branch. The ascent is terminated by the ignition of the degenerate helium core which forms in the center of a star during this period of its evolution. Details of theoretical studies of this process are reviewed by Iben (1974) and Renzini (1977).

Like the main sequence, the RGB of most individual globular clusters is narrow and well defined, placing limits on chemical inhomogeneities (Section 2.3), and on variation in stellar physics among stars on the RGB. A comparison of the RGBs of different clusters reveals that globular clusters of higher metallicity exhibit giant branches that are shallower (in the color–magnitude diagram) and redder than low-metallicity clusters (Sandage and Wallerstein 1960; Sandage and Smith 1966). This is illustrated in Figure 2.2 below. Like stars on the main sequence, higher metallicity giant stars have increased opacity, due to electrons from metals, which allows stars to maintain equilibrium at a lower temperature. The astrophysics of stars on the RGB is complex, and the exact location of the RGB is dependent on mass-loss rates and the details of the convective processes within such stars, often treated in terms of 'mixing-length theory' (see Sweigart and Gross 1978; Suntzeff 1993).

2.1.3   Horizontal Branch

The horizontal branch (shown in Figure 2.1) is composed of stars with helium-burning cores which have evolved off the RGB. The horizontal branch is identified on color–magnitude diagrams as a strip of stars, bluer than the RGB and brighter than the main sequence, which have a range of colors but similar luminosities (thus 'horizontal' on the usual color–magnitude diagram). The presence of RR Lyrae variables within the 'instability strip' or 'RR Lyrae gap' on the horizontal branch is a defining feature of Population II stars. A color–magnitude diagram of the Population I stars of the Galactic disk is free of stars in this region.

The horizontal branch of globular clusters has a particular importance in understanding these systems, as well as shedding light on stellar and Galactic evolution. A simple method of describing the morphology of the horizontal branch is through a measure of the relative numbers of stars blueward and redward of the RR Lyrae gap. One way of quantifying this is through the parameter:

$$C \equiv (B - R)/(B + V + R), \tag{2.1}$$

where B is the number of stars on the horizontal branch on the blue side of the RR Lyrae gap, R the number of stars on the red side, and V is the number of variables on the horizontal branch (Zinn 1986; Lee 1990). In order to explain the detailed location of stars on the horizontal branch, it has been known for some time that mass loss during the earlier RGB phase must be invoked (e.g. Rood 1973).

Observationally, the horizontal branch morphology (i.e., the color distribution of horizontal branch stars) of Milky Way globulars exhibits a broad range (Stetson 1993; Buonanno 1993 and references therein), and is determined by a number of physical effects. In general, the most important parameter is the metallicity of the cluster. Horizontal branch stars of higher metallicity are redder than those of lower metallicity as a result of higher opacity in their envelopes. Metallicity is therefore the 'first parameter' in determining horizontal branch morphology.

Figure 2.2 shows the color–magnitude diagrams of NGC 1904 ([Fe/H] $\simeq -1.7$) and NGC 6637 ([Fe/H] $\simeq -0.6$) and illustrates the effects of metallicity on horizontal branch morphology. NGC 1904 has a large number of stars blueward of the RR Lyrae gap, whereas the metal-rich cluster NGC 6637 has only a short, red horizontal branch. The extended downward tail of the blue horizontal branch of NGC 1904 is *not* exhibited by all metal-poor clusters, as we discuss further below.

Metallicity variations alone can *not* account for some of the observed differences between the horizontal branches of Galactic globular clusters (Sandage and Wallerstein 1960 and many subsequent papers). There are several clusters which have horizontal branches that are too blue for their metallicity, and, more generally, there is no simple one-to-one correlation between metallicity and horizontal branch color. This observation has led to the so-called 'second parameter problem' since it implies that another parameter is influencing the color distribution of horizontal branch stars in globular clusters.

The second parameter problem is illustrated in Figures 2.3 and 2.4. The color–magnitude diagrams of M2 and M3 are compared in Figure 2.3. These clusters have similar metallicities ([Fe/H] $\simeq -1.6$), but their horizontal branch morphologies are clearly different. M2 has a blue horizontal branch exhibiting a long tail, with very

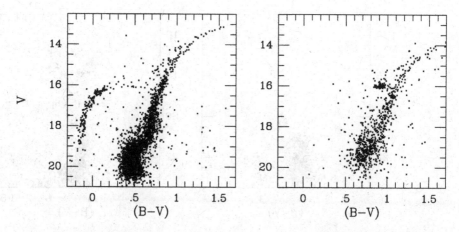

Fig. 2.2. The color–magnitude diagrams of NGC 1904 (left) and NGC 6637 (right) illustrating differences in horizontal branch morphology and the location of the main-sequence turn-off. (From data supplied by R. Buonanno and A. Sarajedini.)

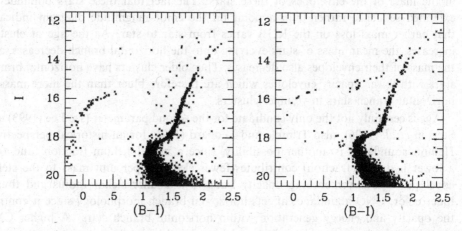

Fig. 2.3. The color–magnitude diagrams of M2 (left) and M3 (right). (From data supplied by P. Stetson.)

few stars redward of the RR Lyrae gap. M3, on the other hand, has a comparable number of stars on each side of the RR Lyrae gap. The clusters in Figure 2.4, Arp 2 and Ruprecht 106, both have comparable metallicities to NCG 1904 shown in Figure 2.2. However, the horizontal branch of Ruprecht 106 is dominated by red stars, whereas Arp 2 has a short, blue horizontal branch. Note that while Arp 2 has a blue horizontal branch, it differs significantly from the much longer horizontal branch of NGC 1904.

Age has long been regarded as a prime candidate for the second parameter. For clusters older than about 10 Gyr, the core mass of stars on the horizontal branch is roughly constant. Thus variations in total stellar mass primarily produce differences

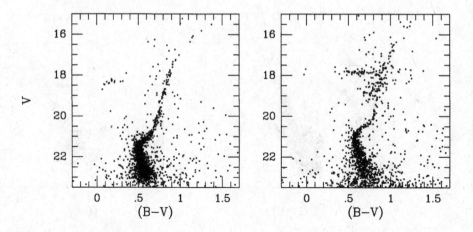

Fig. 2.4. The color–magnitude diagrams of Arp 2 (left) and Ruprecht 106 (right). (From data supplied by R. Buonanno.)

in the mass of the envelopes of these stars. The fact that these stars populate an extended sequence rather than a point on the color–magnitude diagram indicates that earlier mass loss on the RGB varies from star to star. As the age of clusters increases, the mean mass of stars evolving onto the horizontal branch decreases, and the mass of their envelopes also decreases. Thus older clusters have horizontal branch stars with lower opacity envelopes which are therefore bluer than the more massive horizontal branch stars in younger clusters.

Age is certainly not the only candidate for the second parameter (see Lee (1993) and Stetson *et al.* (1996); also Trimble and Leonard (1996) for an historical perspective). Helium abundance is another possibility, since a higher helium fraction (and thus a lower hydrogen fraction) contributes fewer electrons per unit mass to the stellar envelope. The lower ensuing opacity leads to a smaller stellar radius and thus a bluer color. CNO abundance affects horizontal branch morphology since it controls the opacity and energy generation within horizontal branch stars. A higher CNO abundance leads to a redder horizontal branch. A higher stellar core rotation rate may lead to a bluer horizontal branch through its effects on the helium core mass and stellar mass-loss rate. Empirical evidence suggests that globular clusters with high central densities have bluer horizontal branches, perhaps because stellar mass loss in such environments is enhanced (Buonanno *et al.* 1985; Buonanno 1993). Clusters with long blue tails to their horizontal branches, such as NGC 1904 shown in Figure 2.2, also tend to have higher central densities (e.g. Fusi Pecci *et al.* 1993b).

As discussed in Section 2.3 below, it is now generally accepted that age differences exist between some Galactic globular clusters. Thus in some instances, there is little doubt that age is acting as the second parameter in determining differences between horizontal branch morphology. However, there is also considerable evidence that age is not the *only* second parameter, and that variations in some of the quantities listed above also produce an effect (Buonanno 1993). It therefore appears that to understand the full range in horizontal branch morphology, a 'third parameter' (at least!) needs to be included (e.g. Fusi Pecci *et al.* 1993a; van den Bergh and Morris 1993).

Although additional parameters appear to affect horizontal branch morphology, age has long been considered a good candidate to be the dominant second parameter. However, the recent study by Stetson *et al.* (1996) has cast doubt on this idea. These authors have carried out an analysis of age determinations (including some based on new data) which yields smaller age differences between globular clusters than previous studies. While the current state of this field is unsettled, there is reason to be optimistic that the situation will improve through current and future observational programs (see also Chapter 8).

One subtlety in this discussion is the distinction between a 'global' second parameter effect, and second parameters determining horizontal branch morphology in individual globular clusters. The former arises when one treats the Milky Way globulars as a *system*. It has been found that horizontal branch morphology correlates with Galactocentric distance, suggesting that whatever property is driving changes in horizontal branch morphology is also a function of the location of a globular cluster within the Galaxy. Lee *et al.* (1994) conclude that, of the various second parameter candidates, only age differences can reproduce the global second parameter effect *and* successfully reproduce the observed properties of RR Lyraes and the main-sequence turn-off in globular clusters. However, this conclusion has also been challenged by the work of Stetson *et al.* (1996) mentioned above. We discuss this issue further in Chapter 3.

The RR Lyrae stars that are found on the horizontal branch are of considerable importance in globular cluster research. The relationship between their metallicities and absolute luminosities allows them to be used as distance indicators for Milky Way globulars. Reliable distances are clearly essential to obtain other derived properties of globular clusters such as absolute magnitude and physical size (see Chapter 3).

Galactic globular clusters can be separated into two distinct groups, based on the periods of their RR Lyrae stars (Oosterhoff 1939, 1944). Arp (1955) showed that this 'Oosterhoff effect' reflected a difference in the metal abundance of globular clusters. Understanding the effect of metallicity on RR Lyrae luminosity (and thus period) is therefore desirable in order to use these stars as reliable distance indicators. A discussion of physical models for RR Lyrae variablity is beyond the scope of this text, but is discussed by Sandage (1993a) and VandenBerg *et al.* (1996).

2.1.4 *Asymptotic giant branch and beyond*

When the helium in the core of a horizontal branch star is exhausted, the core contracts and helium begins burning in a shell around the core. Beyond this helium shell is the hydrogen-burning shell. The increase in energy generation means that the star ascends the giant branch for the second time. This second ascent is known as the Asymptotic Giant Branch (AGB) phase of stellar evolution. The location of the AGB in the globular cluster color–magnitude diagram is illustrated in Figure 2.1. Cohen (1976) studied AGB stars in globular clusters and discovered that many exhibited significant mass loss in the form of stellar winds. Other peculiarities include helium shell flashes, believed to be a consequence of the helium shell being spatiallly thin, which can cause the star to migrate briefly to the instability strip of the color–magnitude diagram. Like the RGB phase, the details of the astrophysical processes of stars on the AGB are complex and beyond the scope of this book. A review of the topic is provided by Iben and Renzini (1983). More recent developments are described in Johnson and Zuckerman (1989) and Judge and Stencel (1991).

Calculations of the post-AGB phase of stellar evolution are made more difficult by the occurrence of stellar mass loss. However, a low-mass star eventually exhausts its hydrogen and helium shells and loses its extended envelope, ending up as a white dwarf. As discussed in Section 2.5 below, much of the observational work on white dwarfs in globular clusters has been carried out at X-ray wavelengths. However, the refurbished *HST* has also detected white dwarfs in globular clusters (e.g. Richer *et al.* 1995).

2.1.5   *Blue stragglers*

One oddity of the color–magnitude diagram of globular clusters is the presence of 'blue stragglers', indicated in Figure 2.1. The blue stragglers give the appearance of being an extension of the main sequence beyond the turn-off point. Until relatively recently, only a handful of globulars were known to contain such stars, although open clusters in the Milky Way frequently exhibit the phenomenon. However, subsequent observations have revealed that blue stragglers are a common feature of globular clusters as well. These objects are somewhat removed from the main thrust of this book, but for completeness we discuss them briefly here. Full details can be found in Saffer (1993), Stryker (1993), Bailyn (1995) and Leonard (1996a).

Various formation mechanisms for blue stragglers have been proposed, and at present it seems likely that more than one may play a significant role (e.g. Leonard 1989; Sarajedini 1993; Bailyn 1995). Perhaps the simplest explanation of blue stragglers is that they are stars that formed later than the bulk of the stars in a globular cluster (Roberts 1960). Such objects would still sit on the main sequence, beyond the turn-off point of the majority of stars. The primary problem with this picture is the large implied age difference between the blue stragglers and the other stars in an environment which contains little, if any, gas. There does not seem to be any material out of which such late-forming stars could have been produced.

A more promising scenario involves stellar mass transfer between roughly equal-mass components in binary systems (McCrea 1964; Webbink 1979; Iben 1986). The more massive of the two stars will eventually evolve along the subgiant branch, during which time it expands and overflows its Roche lobe. Mass is then transferred from this subgiant to its companion, which is still on the main sequence. (Angular momentum loss in the form of a magnetic wind can accelerate the process by bringing the stars together more rapidly, in which case the primary need not have evolved significantly.) While the subsequent detailed evolution of the system has different routes, one possible end-point is a common-envelope binary in which the components eventually merge to form a single star. Clearly the product of such a merger will have a mass greater than the main-sequence turn-off mass, but not more than twice this mass. It has additionally been proposed that, under certain (probably rare) conditions, helium-enriched material can end up in the envelope of the resulting blue straggler, leading to a luminosity somewhat in excess of that expected for the star's mass (Bailyn 1992).

Another popular picture invokes stellar collisions to produce blue stragglers (Hills and Day 1976). This has the attraction that the high stellar densities in globular clusters make collisions relatively likely. (A simple calculation shows that the typical time for stellar encounters in globular cluster cores is significantly less than the age of globular clusters.) Even in low-density systems, the collision rate may be sufficiently high thanks to close encounters between binary–single and binary–binary systems (Leonard and Fahlman 1991). Early simulations suggested that stars produced through collisions might be well mixed (Benz and Hills 1987). As in the helium-enriched version of the mass-transfer model, the mixing leads to an enhancement of helium in the envelope of the resulting star. However, more recent simulations indicate that mixing may not occur in the stellar collisions model (Lombardi *et al.* 1995).

Of the other models that have been proposed, only two seem consistent with current observations. The first attempts to extend the main-sequence lifetime of some stars through internal mixing, resulting perhaps from rapid rotation or a strong magnetic field (Wheeler 1979). The second is more radical, and supposes that most stars undergo considerable amounts of mass loss. Blue stragglers are stars that have avoided such a fate, possibly because they are slow rotators (Willson *et al.* 1987).

We do not attempt to give a detailed critique of these formation mechanisms and the observations which constrain them. However, it is worth noting that both the collision model and the mass-transfer model probably lead to the production of *some* blue stragglers, simply because both processes are likely to occur in globular clusters. There is evidence that the luminosity function of blue stragglers is different in high-density and low-density globular clusters, as is qualitatively expected in the collision scenario (e.g., Fusi Pecci *et al.* 1992; Ferraro *et al.* 1993). However, it has also been claimed that the detailed form of this luminosity function probably requires the presence of more than one 'type' of blue straggler, the implication being that more than one formation mechanism is at work (Sarajedini 1993; Stryker 1993, and references therein). In M3, for example, blue stragglers in the core of the cluster have a significantly different luminosity function than blue straggglers in the cluster outskirts (Bailyn and Pinsonneault 1995 and references therein). The luminosity function of the inner stragglers is consistent with expectations from the collision scenario, whereas the outer blue stragglers have a luminosity function that supports mass-transfer associated with primordial binaries. Thus a combination of the collision model and the mass-transfer model (including a fraction of objects with enhanced helium) seems consistent with current observations. Based on observations of blue stragglers in the old open cluster M67, Leonard (1996b) has also concluded that more than one blue straggler formation mechanism is at work.

Studies of the variability of blue stragglers in globular clusters has also been used to infer the mechanisms responsible for their formation. It has been found that there are two distinct causes of variability in these stars (see Bailyn 1995 and references therein). Some blue stragglers are SX Phoenicis stars, whose photometric variability is a result of pulsations. However, eclipsing binaries have been discovered by Mateo *et al.* (1990) amongst the blue stragglers of globular clusters. These authors argued that this result suggests that the blue stragglers in NGC 5466 formed through the coalescence of primordial binaries, primarily because the stellar density in this cluster is sufficiently low that stellar collisions are expected to be relatively rare. However, there are other clusters, such as 47 Tuc, where the number and luminosity functions of blue stragglers are more suggestive of a collisional origin (Bailyn 1995 and references therein). These considerations reinforce the view that blue stragglers in globular clusters have more than one formation mechanism.

2.2   Globular cluster ages

The ages of globular clusters has been a topic of great interest for many years. The primary reason for this interest is cosmological: the universe must be older than the objects within it. Milky Way globulars contain the oldest stars for which reliable age estimates are available, and thus provide a lower limit to the age of the universe. Derived globular cluster ages have frequently been comparable to (or greater than) cosmological age determinations, such as those obtained through measurements of the Hubble constant. Such results have been cited as both a success and a serious problem for cosmological models (e.g. Tayler 1986; Sandage 1993b; Bolte and Hogan 1995, and references therein). A comprehensive review of globular cluster ages is provided by VandenBerg *et al.* (1996).

The fundamental method of establishing the absolute age of a globular cluster is to determine the mass of stars at the main-sequence turn-off. The observational parameters that provide the input to this process are the color and apparent magnitude of the turn-off, along with the distance to the cluster and the reddening along the line of sight. This allows a calculation of the absolute magnitude M_V of the main-sequence turn-off. Stellar atmosphere models are then used to determine the total energy produced in the core of the star (the bolometric magnitude) and the effective temperature of the stellar photosphere. Additional input into these models includes metallicity and helium mass fraction. The bolometric magnitude and effective temperature are compared with the results of stellar interiors models which relate these quantities to stellar mass and age.

It has been found that the luminosity of the main-sequence turn-off provides the most reliable estimate of cluster ages. The color of the main-sequence turn-off can also be used, but there are greater uncertainties in stellar models in predicting this color as a function of age (e.g. Bolte and Hogan 1995). The primary source of uncertainty in using the luminosity of the main-sequence turn-off is the distance to a given cluster, although uncertainties in other quantities such as metallicity and elemental abundance ratios must also be taken into account (e.g. Demarque 1980; Wheeler *et al.* 1989; VandenBerg 1992). Nearby subdwarfs, whose distances can be determined to high accuracy through trigonometric parallax, provide standard candles for estimating cluster distances. The absolute magnitude of such stars is a function of color and metallicity, so a number of them are required to provide a complete calibration. The magnitude offset between the observed main sequence of a cluster and the locally calibrated main sequence of subdwarfs gives the distance modulus of the cluster. Over the last few years, the growing observational database of parallaxes, metallicities and colors for nearby subdwarfs has reduced the uncertainty in cluster distances (van Altena *et al.* 1991).

The well-studied globular cluster M92 is an excellent object for age determinations. It has a high galactic latitude, so that interstellar reddening is not a problem, and a low metallicity. The latter is helpful since the stellar models are more reliable at low metallicity. Moreover, in some scenarios for the formation of the Milky Way (see Chapter 3) the most metal-poor globulars are expected to be the oldest. The turn-off magnitude of $M_V = 3.95$ indicates, at the metallicity of M92 ([Fe/H] $= -2.26$), an age of 15.8 Gyr (Bolte and Hogan 1995). Random errors, primarily due to uncertainties in distance and [Fe/H], combine to give a standard deviation in this age of 2.1 Gyr. Model uncertainties are *not* included in the error budget.

An independent check on the distances derived from the subdwarf technique is naturally desirable. The white dwarf cooling sequence is one such alternative. This technique relies on matching local white dwarfs to those observed in globular clusters in much the same way that the subdwarf technique uses nearby subdwarfs as standard candles to estimate the distance to globular clusters. However, the possible systematic problems are different, since the physics of white dwarf cooling is much different than the physics of subdwarf stars. With the refurbished *HST*, it has become technically feasible to observe white dwarfs in nearby globular clusters. The first results from this technique for the globular cluster NGC 6752 have been presented by Renzini *et al.* (1996). They find a distance consistent with other estimates, and derive an age of around 15 Gyr. It should be noted, however, that this result requires an estimate of the masses of the observed white dwarfs, but this estimate is fairly well constrained by independent considerations (see Renzini *et al.* 1996 for a full discussion).

Model uncertainties in the age estimates are more difficult to assess (see VandenBerg *et al.* 1996 for a detailed review). Model stellar isochrones appear to fit globular cluster color–magnitude diagrams very well. As an example, Figure 2.5 shows isochrones of different ages overlayed on the M92 color–magnitude diagram, illustrating the impressive agreement between the stellar models and the observed properties of this cluster. However, the models are dependent to varying degrees on parameters which are not directly observable (e.g. the initial helium abundance). Shi (1995) and Schramm *et al.* (1995) have studied high initial helium abundances in globular clusters, as well as stellar mass loss, and conclude that globular cluster ages could be as low as 10 Gyr, but no lower. However, such low ages are only possible if the initial helium abundance in globular clusters significantly exceeds the current upper observational limits on the primordial cosmological value, thereby requiring helium production prior to the formation of globular clusters. Such a scenario is clearly rather speculative.

A recent attempt to assess the uncertainties in stellar models concludes that the age of the oldest globular clusters lies in the range 11–21 Gyr (Chaboyer 1995). The greatest model uncertainty found is this study is due to limits in our understanding of the convection process in low-mass stars. Helium diffusion has also been studied as a possible source of uncertainty in derived ages (Proffitt and VandenBerg 1991 and references therein). While there is some evidence that such diffusion can reduce inferred ages, Chaboyer *et al.* (1992) conclude the effect is not large.

Globular cluster age determinations are currently an active area of research, both through theoretical and observational developments. Jimenez and collaborators have extended several techniques for determining cluster ages. One of these exploits the stellar luminosity function, which Jimenez and Padoan (1996) argue can provide age estimates which are less sensitive to uncertainties in globular cluster distances and

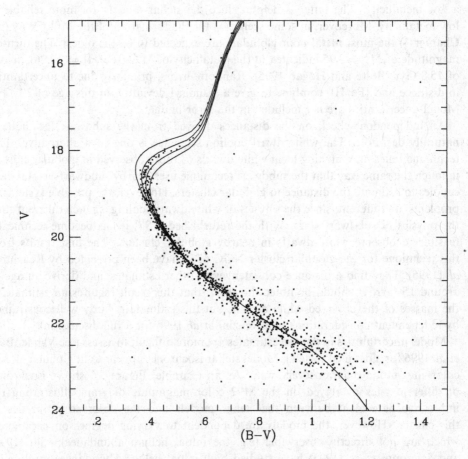

Fig. 2.5. The color–magnitude diagram of M92 with isochrones from Bergbusch and VandenBerg (1992) overlaid. The isochrones represent ages of (left to right): 12, 14, 16 and 18 Gyr. (From data supplied by M. Bolte.)

stellar evolutionary model parameters. The method is based on the fact that the number of main-sequence stars in a bin of fixed luminosity decreases as the stellar population evolves. Star counts on the main sequence and RGB are used to obtain an age estimate for a given cluster. A different approach employed by Jimenez *et al.* (1996) involves determining the mass and mass-loss rate of stars on the red giant branch, the latter being constrained by horizontal branch morphology. For a sample of eight globular clusters, these authors obtain ages for the oldest objects of 13.5 ± 2 Gyr, somewhat lower than more traditional estimates.

Age differences between Milky Way globular clusters are of considerable interest, since they provide constraints on the timescale of the formation of the Galactic halo. Further, relative ages of globular clusters are more reliable than absolute ages, primarily because determinations of age differences are less sensitive to stellar models. As noted in Section 2.1 above, age may be the most important factor in producing horizontal branch differences in clusters of the same metallicity. Comparing clusters of the same metallicity is particularly fruitful, since this eliminates the need for stellar atmosphere models and accounting for the effects of metallicity on the color–magnitude diagram.

Several methods for determining relative ages have been employed (see Bolte 1993 for a detailed discussion). The first widely used technique exploits the luminosity difference between the main-sequence turn-off and the horizontal branch (Iben and Faulkner 1968; Sandage 1982). As discussed in Section 2.1, the core mass of globular cluster stars on the horizontal branch is roughly constant. This translates into a horizontal branch luminosity which is almost independent of cluster age. The main-sequence turn-off magnitude continues to decrease with increasing cluster age, so the magnitude difference between the horizontal branch and the main-sequence turn-off provides an estimate of the age of a globular cluster.

Although this method is conceptually simple, it does suffer from several problems. One serious drawback is that it is difficult to apply the technique to globulars which only have stars blueward of the instability strip. Such clusters are likely to be the oldest and thus of considerable interest when searching for age differences. As Figure 2.1 illustrates, this part of the horizontal branch is not particularly horizontal, making it difficult to pin down its magnitude. In practice, the magnitude difference between the horizontal branch and main-sequence turn-off is usually determined at the color of the turn-off, so for clusters with blue horizontal branches an extrapolation must be made.

A bigger problem is uncertainities in the absolute magnitude $M_V(RR)$ of RR Lyrae stars (and thus the the horizontal branch) and the dependence of this magnitude on metallicity (see also Section 3.1). The magnitude–metallicity relation is usually parameterized in the form:

$$M_V(RR) = c_0[Fe/H] + c_1, \tag{2.2}$$

where c_0 and c_1 are constants. Current estimates put the value of c_0 in the range between 0.1 and 0.3, and there is sufficient uncertainty to prevent a reliable determination of the distribution of globular cluster ages.

A second technique for deriving relative cluster ages involves a detailed comparison of the color–magnitude diagrams of target clusters (see Stetson *et al.* 1989; Bolte 1989; Green and Norris 1990; Richer *et al.* 1996). This method is similar to the absolute age determination described above, but instead of obtaining a distance modulus of a cluster based on the offset between its main sequence and that defined by the local subdwarfs, a relative distance modulus between two clusters is obtained. Interstellar reddening must be taken into account and, for clusters of different metallicity, a further correction must be made based on stellar atmosphere models. Once this is achieved, the difference in the magnitude of the main-sequence turn-off of the two clusters provides an estimate of any age difference. Uncertainties in reddening and metallicity corrections can translate into a significant uncertainty in relative age of around 2 Gyr. In order to minimize this problem, it is profitable to compare clusters of similar metallicity and well-known reddening.

A more recently developed technique relies on the color difference between the main-sequence turn-off and the base of the giant branch (VandenBerg *et al.* 1990; Sarajedini and Demarque 1990). Both these points in the color–magnitude diagram tend to be bluer for younger clusters, but the color variation with age of the main-sequence turn-off is more pronounced. Consequently, older clusters have a smaller color difference between the main-sequence turn-off and the base of the giant branch.

Studies of the relative ages of nearby, metal-poor globulars (including M92 discussed above) indicate an upper limit on any age spread of less than 0.8 Gyr. However, there is also strong evidence for age differences between some globulars. Probably the best-established cases are Ruprecht 106 (Buonanno *et al.* 1990, 1991; Da Costa *et al.* 1992) and Pal 12 (Gratton and Ortolani 1988; Stetson *et al.* 1989), both of which appear to be around 3 Gyr younger than other globulars of similar metallicity. A detailed relative age study of NGC 288 and NGC 362 suggests an age difference between these globulars of around 2–3 Gyr (Bolte 1989; Green and Norris 1990). There are other candidates for younger Milky Way globulars that are similar to Pal 12 in that they are of rather low luminosity. Chaboyer *et al.* (1996) used the magnitude difference between the main-sequence turn-off and the horizontal branch to derive ages for 43 Galactic globular clusters. Within their sample, these authors found an age spread of 5 Gyr.

The interpretation of these results is complicated by the fact that studies have often concentrated on clusters with peculiar color–magnitude diagrams, since they are the best candidates for finding age differences. The current state of this field is consistent with an age distribution in which most Milky Way globulars are uniformally old (with no detectable age spread), but with a few globulars being significantly younger (Bolte 1993 and references therein). We discuss the implications of these findings for the formation of the Milky Way halo in the next chapter.

2.3   Chemical properties

As discussed in Section 2.1, color–magnitude diagrams are sensitive to stellar helium abundance, since an increase in the helium fraction leads to bluer stellar colors. The color dispersion of main-sequence stars at a given magnitude in M92 implies an upper limit on star-to-star variations in the helium abundance of 0.03 (Stetson 1993). This compares to an overall helium abundance of around 0.25.

Similar arguments can be applied to the giant branch of globular cluster color–magnitude diagrams. Here, elements such as Fe, Si and Mg have a significant effect on the opacity of the outer regions of the stars, so that star-to-star variations in these elements produce a dispersion in the color of the giant branch. Observations of the giant branch of NGC 288 imply a dispersion in the abundance of iron-peak elements, [Fe/H], of less than 0.07 dex (Bolte 1992).

Spectroscopy of giants is potentially a more powerful technique for detecting or limiting metallicity variations, although only relatively recently have enough stars per cluster been observed to derive a useful limit on the metallicity dispersion (see Suntzeff 1993 and references therein). These studies support those based on color–magnitude diagrams. Internal homogeneity has been found to apply to globular clusters spanning the full range of metallicity of the Galactic globular cluster system. One Milky Way globular cluster, ω Cen, does *not* exhibit this property (Woolley *et al.* 1966; Smith 1987; Suntzeff and Kraft 1996; Norris *et al.* 1996), showing instead a dispersion in [Fe/H] of 0.30 dex or more (Norris 1980; Mukherjee *et al.* 1992). There have been claims that M22 has similar, but smaller, [Fe/H] variations (Lloyd Evans 1975; Norris and Freeman 1983), but the observational situation is complicated by differential reddening across the cluster. At present, the existence of an [Fe/H] spread in M22 remains controversial (e.g. Lehnert *et al.* 1991; Anthony-Twarog *et al.* 1995). There are one or two other Milky Way globular clusters that are candidates for [Fe/H] variations (e.g. Langer *et al.* 1993), but the vast majority of clusters clearly have very little star-to-star spread in the abundance of iron-peak elements.

One reason that metallicity homogeneity within globular clusters is so remarkable is that it sets them apart from other stellar systems. Galaxies show significant star-to-star metallicity variations: in the Milky Way, for example, there are some stars with metallicities below 10^{-4} of the solar value (e.g. Preston *et al.* 1991). The local dwarf spheroidal galaxies are of particular interest in this regard, since they have comparable integrated luminosities to the brightest globular clusters, but have metallicity dispersions of a few tenths of a dex (e.g. Suntzeff 1993). This suggests that star formation in globular clusters occurred in a different fashion to star formation in other stellar systems. It is also interesting that metallicity differences between different Milky Way halo globular clusters are of the same order as the differences between halo stars, although the metal-poor tail of the distribution of halo field stars and globular clusters differs (see Chapter 3).

Despite the homogeneity in [Fe/H] within individual clusters, there are significant star-to-star variations in the abundance of certain elements. The best known are the CN variations found among globular cluster stars (Smith 1987; Suntzeff 1993, and references therein). The central question posed by these variations is whether they reflect star-to-star differences built in when a globular cluster formed, or whether they result from stellar evolutionary processes. Stellar evolution is expected to produce some abundance variations between stars at different evolutionary phases. Standard theory predicts that once a star reaches the giant branch its convective envelope is sufficiently deep that it mixes material that has undergone CNO processing up to the surface levels of the star. This general picture has observational support. In many clusters, it has been found that [C/H] decreases as stars ascend the giant branch. Further, stars on the upper giant branches of globular clusters have large [N/Fe] ratios. These findings are consistent with the idea that CN-processed material is dredged up to the surface layers of red giants (Smith 1987). Unfortunately, the observed decline in $^{12}C/^{13}C$ and [C/H] with evolutionary stage on the giant branch is not in quantitative agreement with standard models. Specifically, observations require more mixing than is theoretically predicted (VandenBerg 1992).

An interesting aspect of these studies is that carbon depletion is weaker in metal-rich clusters, implying that deep mixing is more efficient in low-metallicity stars. The meridional mixing model predicts such a phenomenon (Sweigart and Mengel 1979). Details of this picture are beyond the scope of this book, but essentially the model predicts that the regions in which CN and ON processing occurs extend further from the hydrogen-burning shell in low-metallicity stars. This leads to more CN-processed material getting mixed into the surface layers of such stars. One problem with this model is that it fails to reproduce C and N variations low on the giant branch. It is worth noting in this context that CN variations are present in *main-sequence* stars in 47 Tuc and NGC 6752 (Bell *et al.* 1983; Suntzeff 1989; Briley *et al.* 1991), and *no* theoretical models predict that mixing can dredge up CN-processed material when stars are still on the main sequence.

The CN variations within a given cluster exhibit the surprising property of being bimodal (e.g. Norris and Smith 1981). That is, at a given magnitude, stars have either strong or weak CN bands, with few stars having intermediate values. It is not clear at present whether this bimodality has a primordial origin, or whether it results from mixing on the red giant branch. However, the ratio of CN-rich to CN-poor stars shows considerable variation from one cluster to another. It has been argued that some global characteristic of globular clusters is influencing this ratio (Suntzeff 1993, and references therein). This view may be supported by the observation that halo field stars are generally very weak in CN. In other words, Population II giants outside of globular clusters do not include the CN-rich population found in some globulars.

Other elements also exhibit abundance variations in globular clusters. The [O/Fe] ratio in globular cluster giants sometimes exhibits considerable depletion relative to halo field stars, which are enhanced by a factor of several relative to the solar abundance (e.g. Wheeler *et al.* 1989). The degree of depletion varies among globular clusters, with some clusters having many stars with severely depleted [O/Fe] and others showing much less pronounced depletion (e.g. Kraft *et al.* 1993 and references therein). This again suggests that a global characteristic of globular clusters may be driving the degree of depletion. Low [O/Fe] values can be produced by CNO-processing. If this is the only mechanism at work, the combined abundance of C, N and O should be a constant within a given cluster (no star-to-star variations). At present, there does not appear to be clear evidence for or against this prediction.

Finally, Al and Na (Cohen 1978; Cottrell and Da Costa 1981) and Mg (Shetrone 1996) show variations which have yet to be explained in terms of evolutionary processes and which have been interpreted as indicating a primordial star-to-star variation. In globular cluster giants, Al and Na are correlated with each other and anti-correlated with [O/Fe] (e.g. Sneden *et al.* 1992; Kraft *et al.* 1993). Relative to halo field stars, Al and Na tend to be overabundant in globular clusters and show wide variations. If these Al and Na variations *are* primordial, their anti-correlation with [O/Fe] suggests that CNO variations are also primordial. It should be stressed that there are models that attempt to explain Al and Na variations in terms of non-primordial processes (Langer and Hoffman 1995, and references therein). Currently, there does not appear to be a consensus on whether these variations have a primordial or evolutionary origin.

For the main thrust of this book and as clues to the globular cluster formation process, two issues revealed by abundance studies stand out. The first is that the chemical *homogeneity* of iron-peak elements in globular clusters is a unique feature of these objects not shared by other stellar systems. The second is the possibility that some global property of globular clusters may be influencing some of the elemental abundances of their constituent stars.

2.4   Stellar luminosity functions

A complementary method of studying the stellar content of a globular cluster is through its luminosity function (see Fahlman 1993, and references therein). Instead of describing the distribution of stars in terms of color and magnitude, the luminosity function quantifies the relative number of stars at different magnitudes. The differential luminosity function, $n(m)$, is usually defined as the number of stars per unit magnitude, m, through the expression:

$$dN = n(m)dm. \tag{2.3}$$

While this expression is general, studies of globular clusters have the great advantage that all the stars in a given cluster are at roughly the same distance. Thus the number of stars as a function of absolute magnitude, M, has the same form as the observed luminosity function.

Deriving the luminosity function of globular clusters is of interest for many reasons, most notably because it is a critical step in determining the mass function. The mass function of stars in globular clusters reflects the initial mass function with which stars formed and is thus of importance for formation models. The mass function may also be affected over time by mass segregation in globular clusters (discussed in Section 2.6), which leads to the preferential removal of low-mass stars, potentially flattening the slope of the mass function. The mass function also has significant implications for dynamical models of globular clusters, as discussed in Section 2.6.

Because of the great age of Galactic globular clusters, only stars with masses of about 0.8 $M_\odot$ or less are on the main sequence. As more massive stars evolve off the main sequence, they move rapidly through the various stages of post-main-sequence evolution described in Section 2.1, finally becoming stellar remnants. Therefore, there is no direct evidence about the initial mass function for stars more massive than around 0.8 $M_\odot$ in Galactic globular clusters. However, the short post-main-sequence lifetimes imply that the number of stars at a given luminosity is directly proportional to the lifetime spent at that luminosity (see Renzini and Fusi Pecci 1988). Thus the bright end of the luminosity function in globular clusters provides a useful probe of post-main-sequence evolution.

For stars below the main-sequence turn-off, the mass function is directly related to the luminosity function:

$$n(m)dm = n(\mathscr{M})d\mathscr{M}, \tag{2.4}$$

where $\mathscr{M}$ is the stellar mass. The stellar mass function was first parameterized by Salpeter (1955) using the power-law form

$$n(\mathscr{M}) = A\mathscr{M}^{(-1+x)}, \tag{2.5}$$

where x is known as the spectral index of the mass function and A is a normalization constant. In practice, stellar mass functions appear to have a varying value of x as a function of $\mathscr{M}$. These variations are often accommodated by a piece-wise definition of the mass function, with x having different constant values over specific mass ranges (e.g. Scalo 1986).

There are two primary challenges for deriving the mass function below the main-sequence turn-off: determining the luminosity function of faint, low-mass stars in crowded regions, and the uncertain relationship between luminosity and mass. An additional complicating factor is that the mass function need not be constant with radial distance from the cluster center. In fact, mass segregation (see Section 2.6) is both expected and observed. Further, mass segregation and other dynamical effects can produce evolution in the mass function, with the degree of evolution depending on the global properties of the cluster.

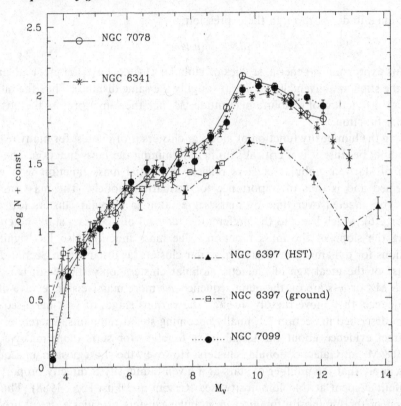

Fig. 2.6. *HST V*-band luminosity functions for NGC 6397, M15, M30, and M92 from Piotto *et al.* (1997). The luminosity function for NGC 6397 has been extended up to the main-sequence turn-off using ground-based data. (Figure supplied by I. King.)

Because there is no substitute for better data, this field has been and will continue to be transformed by *HST* observations. The observational techniques involved in recovering true luminosity functions from deep images of globular clusters diagrams are described by Stetson (1990), Fahlman (1993), and Cool *et al.* (1996). The deepest pre-*HST* observations suggested that the luminosity functions of globular clusters had a variety of slopes at the faint end, with some rising rapidly at the faintest levels, down to masses of about 0.15 $M_\odot$ (Richer and Fahlman 1992). However, *HST* data indicate that, at the lowest luminosities, these analyses are unreliable. Current *HST* data indicate that the luminosity function rises to about 0.2 $M_\odot$, then declines (Piotto *et al.* 1997; de Marchi and Paresce 1995a,b; Elson *et al.* 1995). These results probably rule out a significant population of objects in globular clusters with masses below about 0.1 $M_\odot$.

Examples of luminosity functions for four globular clusters are presented in Figure 2.6. The data combine *HST* and ground-based observations. One point of considerable interest is that NGC 6397 exhibits a significantly shallower luminosity function slope than the other three clusters. The most likely explanation for this difference is that NGC 6397 has experienced a greater degree of mass loss than the other clusters, such that low-mass stars have been preferentially removed. As discussed in Section 2.6 below, mass segregation in globular clusters leads to low-mass stars occupying the outer regions of a cluster. Such stars are more vulnerable to removal from the cluster through various dynamical processes such as shocks resulting from crossing the Milky Way disk (see Section 3.5). Based on its location in the Galaxy and proper motion studies of its stars, NGC 6397 has experienced more such disk-crossings than the other three clusters illustrated in Figure 2.6. Thus the shallower luminosity function of this cluster provides strong evidence that dynamical evolution, and evolution of the luminosity function, has taken place. The details of the observations and analysis of these four luminosity functions are given by Piotto *et al.* (1997).

A major issue in these studies is the conversion of luminosity to stellar mass. This conversion is particularly difficult to calculate for the low-mass cluster stars, as there are still significant uncertainties in both the stellar structure and the atmospheres of these cool stars. This field is currently active, and luminosity to mass conversions have been published by a number of groups (e.g. Bergbusch and VandenBerg 1992; Baraffe *et al.* 1995; D'Antona and Mazzitelli 1996; Alexander *et al.* 1996). Many of these models can fit the prominent features observed in the faint luminosity functions, although significant differences between the different models remain, and a consensus on the best mass-to-light conversion for very low-mass stars has not yet been reached.

Attempts to find correlations between the mass function slope x and other globular cluster properties have mostly produced negative results. However, these studies have been limited to a fairly narrow range of masses, and new *HST* results may bring about changes in our understanding. One possible exception to the generally null results is a correlation between x and the position of the globular cluster within the Galaxy. We return to this topic in Chapter 3 since such a correlation is more usefully examined in terms of the Milky Way globular cluster system as a whole.

2.5 Binaries and stellar remnants

The great age of Milky Way globulars means that a significant fraction of their original stars has evolved off the main sequence. The final fate of these stars includes the formation of white dwarfs, neutron stars, and possibly black holes. The high central densities of globular clusters are also relevant, since they increase the probability that binary systems of compact objects form. We will not treat formation models of such objects, nor the astrophysics of these systems, in detail. For an overview of these topics, the reader is referred to Bailyn (1995), Grindlay (1993), Phinney (1993), and Meylan and Heggie (1996). Instead, we give a brief description of the observations of binaries and stellar remnants in globular clusters.

Globular clusters are excellent laboratories for observational studies of the late stages of stellar evolution and the formation of compact objects because of the large number of stellar remnants expected. Further, these remnants are all located at the same distance and come from stars which formed at the same time. The primary difficulty of studying objects in globular clusters is crowding, which can make imaging of individual objects difficult, and can complicate analyses of the dynamical signatures of objects in binary systems.

The study of white dwarfs has advanced tremendously through imaging with the refurbished *HST*. White dwarf sequences are visible on color–magnitude diagrams of M4 (Richer *et al.* 1995), NGC 6397 (Cool *et al.* 1996; Paresce *et al.* 1995) and NGC 6752 (Renzini *et al.* 1996). At this point, the primary conclusions are that white dwarfs exist in globular clusters, and that the numbers of objects which have recently evolved to the white dwarf stage are generally consistent with expectations. Further progress is expected from current and future *HST* programs.

Searches for accreting white dwarfs (cataclysmic variables or CVs) have also been undertaken. Optical identifications are challenging, but several good candidates have been identified, particularly through *HST* imaging (e.g. Paresce and DeMarchi 1994; Cool *et al.* 1995). X-ray observations are a useful way to search for CVs, since CVs in the field are known to be X-ray emitters, and few other stars in globular clusters are expected to be X-ray sources. *Einstein* observations first revealed a population of low-luminosity X-ray sources in globulars which were conjectured to be CVs (Hertz and Grindlay 1983a,b). The total X-ray luminosity from these low-luminosity sources appears to be consistent with theoretical expectations (Di Stefano and Rappaport 1994). Observations made at much higher resolution with the *ROSAT* high-resolution imager have confirmed the existence of low-luminosity X-ray sources in globular clusters, and allowed a more reliable identification of sources in globular clusters (see Bailyn 1995).

Bright X-ray sources in globular clusters are generally accepted to be produced by accreting neutron stars. When a neutron star captures a companion, the accretion of material produces X-ray emission and the object is observed as a low-mass X-ray binary. While such observations provided the first evidence for neutron stars in globulars, the finding has been confirmed by the presence of millisecond pulsars in globular clusters. The large numbers of these sources have led to an increase in the estimated number of neutron stars in these systems (Phinney and Kulkarni 1994).

At present, there is no compelling evidence for the presence of black holes in globular clusters. It has been suggested that the surface brightness profiles in some post-core-collapse clusters (see Section 2.6 below) may indicate the presence of a central black hole, but this is currently a matter of debate. If black holes *do* exist in globulars, they are, in principle, detectable through their X-ray signatures.

Observations of CVs provide information on binaries involving white dwarfs, but there has also been considerable work devoted to other types of binary in globular clusters (see Meylan and Heggie (1996) for a comprehensive review). One of the primary goals of these studies is to establish whether the frequency of binaries in globular clusters differs from that amongst field stars. The high stellar densities characteristic of globular clusters promote the formation of binaries through stellar capture, thereby increasing the number of binaries above the primordial value. However, the high densities also lead to collisions and interactions that can disrupt binaries (Meylan and Heggie 1996, and references therein).

Various observational strategies have been employed to detect binaries in globular clusters. Photometric studies have concentrated on eclipsing binaries. Since the discovery of an eclipsing binary in ω Cen by Niss *et al.* (1978), several other programs have detected such systems (e.g. Mateo *et al.* 1990; Yan and Mateo 1994; Rubenstein and Bailyn 1996, and references therein). Ground-based studies are generally restricted to loose globular clusters and the outer parts of more compact clusters because of the problems of crowding. Observations with *HST* have extended such searches into the cores of concentrated clusters (e.g. Gilliland *et al.* 1995; Edmonds *et al.* 1996).

While studies of eclipsing binaries are ideal for detecting short-period binaries, longer-period systems can be found by measuring stellar radial velocities. Early work suggested that the frequency of long-period binaries in globular clusters was low (Gunn and Griffin 1979), but subsequent studies have revised this conclusion (Côté *et al.* 1994; Mayor *et al.* 1996).

There are several problems in using these observational results to compare the frequency of binarism in globular clusters to that in the field. The current database for globular clusters is still relatively small and is comprised of surveys adopting different strategies and probing different period ranges. Further, it appears that the binary fraction varies from one globular cluster to another (Meylan and Heggie 1996 and references therein). Thus most estimates of the binary fraction in globular clusters rely on comparisons between observations and Monte Carlo simulations. This allows an estimate of the number of binaries expected to be detected by a given survey as a function of the true binary fraction.

Unfortunately, there is currently little consensus on whether the primordial binary fraction in globular clusters differs from that of field stars in the solar neighborhood and open clusters. Current results on ω Cen point to a lower binary frequency in this cluster than in the field (Côté *et al.* 1996; Mayor *et al.* 1996), but studies of other globular clusters conclude that the frequency of binaries is similar to that in field stars. It is currently unclear whether this is a consequence of variations in the binary frequency between globular clusters, or whether the discrepancy lies in the interpretation of the simulations from which the binary fraction is derived.

2.6 Dynamical and structural properties

2.6.1 Surface brightness profiles

In the image of the nearby globular cluster M92 shown in Figure 1.1, one key feature is immediately apparent: a significant fraction of the stars is concentrated into a small central 'core' with the remainder of the stars forming a more tenuous distribution. The high stellar density in the central regions of globular clusters is one of the properties which differentiates them from lower-density (and usually lower-mass) open clusters found predominantly in the disk of the Galaxy. However, there is also a large range of central densities within the globular cluster population.

The surface brightness profile of most globular clusters is well fit by a King model; specifically the density profile which results from the following distribution function:

$$f(E) = \begin{cases} 0 & E > E_0 \\ K\left[e^{-\beta(E-E_0)} - 1\right] & E < E_0 \end{cases} \tag{2.6}$$

(King 1966; also Michie 1963), where K, β, and E_0 are constants, $E = (1/2)v^2 + \phi(r)$, $\phi(r)$ is the potential, and $f(E)d\mathbf{r}d\mathbf{v}$ is the mass within $d\mathbf{r}d\mathbf{v}$. The dynamical motivation for models of this type is discussed in the following subsection. Here, we focus on the density distribution and its applicability to observed profiles of globular clusters. For

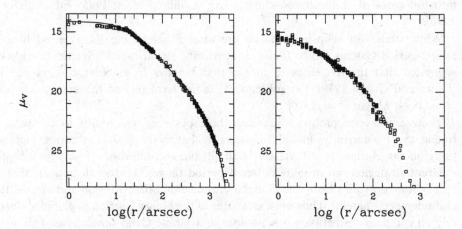

Fig. 2.7. *V*-band surface brightness profiles of 47 Tuc (left) and NGC 7099 (right). Overlaid on the observed profile of 47 Tuc is a King (1966) model with $c = 2.04$ and $\log(r_c/\text{arcsec}) = 1.35$. (From data supplied by S. Trager.)

this model, the core radius is defined as:

$$r_c = \sqrt{\frac{9}{4\pi G\rho_0 \beta}} \tag{2.7}$$

where ρ_0 is the central density. The tidal radius r_t is defined as the point at which $\phi = 0$, and the concentration, c, of such a distribution is defined to be:

$$c \equiv \log(r_t/r_c). \tag{2.8}$$

Observationally, globular clusters are sometimes described in terms of the half-light radius, $r_{1/2}$. This is the radius containing half of the projected integrated light of the cluster. Although the core radius is often numerically similar to the half-light radius, the two quantities are not identical.

King models form a sequence of models that can be specified by c. At any value of c there is a set of models generated by different values of r_c. In many calculations, it is convenient to characterize the concentration through a dimensionless parameter $W_0 = -\beta\phi_0$, where ϕ_0 is the value of the potential at the cluster center. All dimensionless quantities, such as the ratio of the central to the mean density, are determined by W_0. In Figure 2.7 we show a King model fit to the surface brightness profile of the globular cluster 47 Tuc (NGC 104). Also shown is the post-core-collapse cluster NGC 7099 which shows a central cusp rather than the flat core characteristic of the surface brightness profiles of most globular clusters (Djorgovski and King 1986 and references therein). We discuss the phenomenon of core collapse in the next subsection and in Section 3.5.

For clarity and completeness, we also note that there are two other models which have at various times been called King models in the literature (see Richstone and Tremaine 1986). One of these is an empirical fit to observed surface brightness profiles of clusters introduced by King (1962). As discussed in King (1966), these original empirical fits are similar to those derived from the distribution given above. The King (1966) model supercedes the earlier work, as it fits the surface brightness profiles equally well and has a dynamical basis. Another model which is sometimes called a King model (or, more commonly, a modified Hubble profile) has the form: $I(r) = I_0(1 + [r/r_c])^{-2}$ (Rood *et al.* 1972), where I_0 is the central surface brightness. This profile does *not* provide a particularly good fit to most globular clusters.

All these models assume that globular clusters are spherical systems. Observations of Milky Way globulars have found this to be a good approximation, yielding a mean axial ratio of 0.92 ± 0.01 (e.g. White and Shawl 1987).

2.6.2 *Dynamical properties*

There is a vast literature on the topic of the dynamics of stellar systems in general, and globular clusters in particular. Recent reviews include those of Elson *et al.* (1987) and Meylan and Heggie (1996), along with several papers in the conference proceedings dedicated to this topic (Djorgovski and Meylan 1993). Spitzer's (1987) book on the subject provides a comprehensive overview.

We are primarily concerned with aspects of globular cluster dynamics that relate to the formation and evolution of these objects, as well as the use of dynamical modeling to better understand the properties of individual globular clusters. In this section we restrict our attention to internal dynamical processes that are mostly independent of the location of a globular cluster within the Milky Way. The principal dynamical effects that lead to significant globular cluster evolution, and possibly destruction, are dependent on the Galactic tidal field and thus the position of globular clusters within the Galaxy. These effects are more naturally considered in the context of the Milky Way globular cluster *system* and are discussed in Section 3.5.

For many Milky Way globular clusters, the only available mass estimates are obtained by combining an observed cluster luminosity with an assumed mean mass-to-light ratio for the cluster stars. However, in some cases, the central line-of-sight velocity dispersion, σ_0, has been determined, either through the broadening of spectral lines in integrated light spectra, or through the measurement of the individual velocities of a number of stars in the cluster. A dynamical mass estimate can be obtained by combining σ_0 with I_0 and $r_{1/2}$ defined in the previous subsection. The mass-to-light ratio is related to these quantities through the expression:

$$\frac{M}{L} = \eta \frac{9\sigma_0^2}{2\pi G I_0 r_{1/2}} \tag{2.9}$$

where η is a dimensionless constant. Obtaining a mass-to-light ratio in this manner is often referred to as 'core fitting' and was first described in detail by Rood *et al.* (1972).

For the King (1966) model, numerical studies have shown that η has a slight dependence on W_0, but is close to unity for all reasonable models (Richstone and Tremaine 1986). The mass-to-light ratio, along with the integrated luminosity of a cluster, gives an estimate of the cluster mass. This technique has the advantage of utilizing observational quantities which are readily accessible. However, it has two drawbacks which originate in its use of the central velocity dispersion. Firstly, it assumes isotropic orbits. Secondly, it assumes that the mass-to-light ratio is independent of radius. Neither assumption is likely to be strictly true, and there are theoretical reasons to expect the latter not to be the case. The many two-body encounters between stars in globular clusters tend to produce energy equipartition between the stars (all stars having the same kinetic energy). As a result, lower-mass stars acquire higher velocities and preferentially occupy the outer regions of clusters. This process is known as mass segregation and has a number of important implications for the dynamical evolution of globular clusters (see Section 3.5). Since low-mass stars have high mass-to-light ratios, it is likely that the mass-to-light ratio in clusters increases with increasing cluster radius.

One attempt to overcome these problems is the use of King–Michie dynamical models (Michie 1963). These models are similar to the King (1966) model in that they involve lowered Maxwellian energy distributions, but they include the possibility of velocity anisotropy. To account for the different velocity distributions of stars with different masses, it is usual to consider multi-component models in which the spectrum of stellar masses is approximated by several populations, each with a different fixed mass. Stellar remnants such as white dwarfs can also be included as a separate population. The general scheme is described by Gunn and Griffin (1979). Instead of the King (1966) distribution, one has an energy–angular momentum distribution function of the form:

$$f_i(E, J) \propto (e^{-A_i E} - 1)e^{-\beta J^2}. \tag{2.10}$$

Here the subscript i identifies the individual populations and A_i is proportional to the mean mass of stars in the ith population. The relative number of stars in each mass bin is determined by the mass function index x.

In order to obtain a mass-to-light ratio, a large number of models are constructed. Each model is described by the value of x along with values for $r_{1/2}$, W_0, a scale velocity, and the anisotropy radius (see, for example, Meylan (1987, 1988) for a full discussion). An acceptable model must fit both the observed surface brightness profile and the radial velocities of stars in the globular cluster. The method yields not only an estimate of the mass-to-light ratio (and thus the mass), but also the mass function index x and the degree of anisotropy in the stellar velocities (e.g. Meylan 1989). Notably, mass-to-light ratios obtained in this way are typically around 50% higher than those obtained from single-component King models (see Pryor and Meylan 1993; Meylan and Pryor 1993).

More recently, interest has spread to the use of non-parametric methods of determining dynamical and structural properties of stellar systems including globular clusters (e.g. Dejonghe and Merritt 1992; Merritt 1993a,b; Merritt and Saha 1993; Merritt and Tremblay 1994; Gebhardt and Fischer 1995). The primary advantage of such an approach is that it avoids potential biases introduced by assuming a given parametric form to describe the properties of a stellar system. The only drawback is that in order to obtain dynamical properties of globular clusters a large number of radial velocities are typically required. However, this is becoming less of a concern with improvements in multi-object spectrographs. An interesting example of the application of a non-parametric technique to globular clusters is provided by Gebhardt and Fischer (1995). Their technique employs radial velocities and surface brightness profiles and yields mass-to-light ratios (as a function of radius) and stellar mass function indices, as well as other information about the gravitational potential of the clusters in their study.

As mentioned above, the globular clusters with power-law cusps in their surface brightness profiles are believed to have suffered core collapse. This dynamical process involves the migration of stars from the globular cluster core to the outer regions of the cluster, leading to a loss of energy from the core and subsequent core contraction. The phenomenon is sometimes referred to as the gravothermal instability. While this process will occur for an isolated cluster removed from any surrounding gravitational field, studies have shown that the presence of a tidal field tends to accelerate the collapse. Consequently, the position of a globular cluster within the Milky Way is important in determining whether core collapse has occurred by the present epoch. We therefore discuss this topic further in Chapter 3.

2.6.3 Dark matter in globular clusters

One of the important questions that can be addressed by dynamical studies of globular clusters is whether these systems contain significant quantities of dark matter (Heggie and Hut 1996). The presence of stellar remnants implies that, in a limited sense, some dark matter is found in globular clusters, but of more interest is the possibility that globular clusters are surrounded by dark matter halos. There is evidence that galaxies of all morphological types are surrounded by such dark halos (e.g. Ashman 1992), so it is of interest to know whether globular clusters share this property. Perhaps more importantly, the presence or absence of dark matter halos can be used to rule out classes of globular cluster formation models (Chapter 7).

Typical mass-to-light ratios derived for globular clusters using the methods described above are $(M/L)_V \simeq 2$. Such values are consistent with that expected for an evolved stellar population without any dark matter, and in fact are somewhat lower than those predicted by theoretical models of stellar populations with standard initial mass functions (e.g. Worthey 1994). However, most of these estimates are based on velocity measurements in the inner regions of globular clusters. There are two reasons to suppose that any putative dark matter in globulars is likely to be more extended than the visible stars. First, if globular cluster dark halos are similar to those surrounding galaxies, then, like galaxies, one would expect the dark matter to be dynamically significant only at large distances from the centers of globular clusters. Second, if globular clusters initially contain a substantial proportion of low-mass stars or brown dwarfs, mass segregation will lead to these high-M/L objects occupying the outer regions of globulars. Thus, as is the case in other stellar systems, it is difficult to obtain conclusive evidence for or against dark matter using only the motions of the visible stars (Heggie and Hut 1996; Taillet *et al.* 1996).

The limitations of traditional mass-modeling techniques have led to other methods of searching for dark matter. Leonard *et al.* (1992) used proper motion data as well as radial velocities of stars in M13 and concluded that about half the cluster mass may be contained in low-mass stars and brown dwarfs. A more recent technique employed by Moore (1996), however, places strong constraints on the amount of dark matter in the studied clusters. Moore (1996) has made use of observations of tidal tails of stars (Irwin and Hatzidimitriou 1993; Grillmair *et al.* 1995) that are in the process of being stripped from globular clusters by the tidal field of the Galaxy.

The basis of the technique is conceptually simple: if globular clusters have extended dark matter halos, it is less likely that visible stars will be stripped to produce tidal tails. Moore (1996) models the process in some detail and concludes, based on the observations mentioned above, that the global mass-to-light ratios of globular clusters are unlikely to exceed around 2.5. This value is consistent with that of core-fitting analyses and can be accounted for entirely by an old population of visible stars. This study provides the strongest evidence against the presence of dark matter in globular clusters.

2.7   Are Milky Way globulars typical?

We have less detailed information about individual globular clusters in external galaxies. Indeed, beyond the Local Group, current technology severely limits any information on the internal properties of globular clusters. On the whole, most evidence suggests that the Milky Way globular clusters are fairly typical, but there are some important differences.

Shape and structural information on extragalactic globular clusters is only available for the nearest systems. There is evidence that globulars in the LMC are more flattened than their Milky Way counterparts. Globular clusters in M31, on the other hand, seem to be more like Milky Way globular clusters in this regard (Spassova *et al.* 1988; Lupton 1989). While most Milky Way globular clusters are metal-poor, some extragalactic globulars appear to have metallicities well in excess of solar values. Finally, while the great age of Milky Way globulars is one of their defining features, there is mounting evidence for young globular clusters in certain environments. These issues and their significance for models of galaxy and globular cluster formation are discussed in depth in the following chapters.

3

The Galactic globular cluster system

The globular cluster system of the Milky Way consists of over 150 known members. It is likely that not all the clusters have been detected, primarily because of obscuration by the Galactic bulge. The best estimate for the total population is around 180 objects. The system is centrally concentrated, with roughly half of the globular clusters residing within about 5 kpc of the Galactic center. However, the most remote clusters extend to beyond 100 kpc from the center of the Galaxy. The system is more usefully considered as two or more distinct subsystems. The majority of globular clusters form a roughly spherical, metal-poor, halo distribution. Recent evidence suggests that this halo population may itself be a composite system. A smaller number of globular clusters are relatively metal-rich and have the spatial and kinematic characteristics of a thick disk population. It has also been suggested that these metal-rich clusters can more properly be regarded as belonging to the bulge population of the Milky Way.

In this chapter, we examine the range of globular cluster properties, correlations between these properties, and how the characteristics of globular clusters vary with position within the Milky Way. We describe the evidence that has led to the separation of Milky Way globular clusters into distinct subsystems. The dynamical evolution of the Galactic globular cluster system and constraints on the properties of the initial Milky Way globular cluster system are also discussed. Finally, we consider the important question of what the Milky Way globular cluster system reveals about the formation and evolution of the Galaxy.

3.1 Distances to Galactic globular clusters

Many of the observed properties of globular clusters, such as luminosities and radii, are sensitive to the measured *distance* to a given cluster. We therefore begin this chapter by discussing how globular cluster distances are measured.

As mentioned in Section 2.2, globular cluster distances can be estimated by comparing the magnitude of stars on the main sequence to a sample of subdwarfs with distances determined via parallax. The difficulty with this technique is that it requries fairly deep photometry to reach sufficiently far down the main sequence of a given globular cluster. This problem obviously becomes more severe for more distant clusters. In practice, the majority of globular cluster distances are currently obtained by measuring the apparent magnitude of stars on the horizontal branch at the position of the RR Lyrae variable gap. The absolute magnitude of RR Lyraes at the cluster metallicity is then used to obtain the distance modulus of the cluster. Since the horizontal branch is significantly brighter than the lower main sequence, this technique is easy to implement. Unfortunately, there is some controversy concerning the calibration of the RR Lyrae absolute magnitude and its dependence on metallicity (see Sections 2.1 and 2.2; a detailed historical perspective and discussion of this question is provided by Jones *et al.* (1992), Sandage and Tammann (1995), and VandenBerg *et al.* (1996)).

In this book we use the RR Lyrae absolute magnitude calibration employed by Harris *et al.* (1991):

$$M_V(RR) = 0.20[Fe/H] + 1.00. \tag{3.1}$$

This choice is not intended to indicate any preference for a particular calibration and simply reflects the need to adopt a consistent distance scale in order to obtain various globular cluster properties. However, this relation, with slight alterations, has been used by the majority of recent studies. For example, Jones *et al.* (1992) quote $M_V(RR) = 0.16(\pm 0.03)[Fe/H] + 1.02(\pm 0.03)$. Using this relation, Cohen (1992) has noted that RR Lyrae distances to the globular clusters M92 and M5 agree well with those determined using main-sequence fitting.

Sandage and Tammann (1995) present several arguments suggesting that this calibration needs to be revised. Foremost among these is Walker's (1992a,b) finding that, if a calibration similar to that in equation (3.1) is used, the RR Lyrae distance modulus to the LMC differs from the Cepheid distance modulus. Sandage and Tammann (1995) argue that the RR Lyrae calibration is given by:

$$M_V(RR) = 0.30[Fe/H] + 0.94. \tag{3.2}$$

For the metallicity range of most Milky Way globular clusters, this implies that RR Lyraes are around 0.2–0.3 magnitudes brighter than the calibration used by Harris *et al.* (1991). If the Milky Way RR Lyrae calibration is *not* in error, solutions to the discrepancy found by Walker (1992a,b) include a metallicity-dependent metallicity–luminosity relationship for Cepheids (van den Bergh 1995a), or a difference in the RR Lyrae period–luminosity relation between the LMC and the Milky Way (van den Bergh 1995b).

While there are reasonable arguments for both these calibrations, the most recent results seem to favor the calibration in equation (3.1). Of particular note are measurements of RR Lyrae distances to globular clusters in M31 where all clusters are at roughly the same distance (Ajhar *et al.* 1996; Fusi Pecci *et al.* 1996). These studies are consistent with the calibration given in equation (3.1), but are only marginally compatible with the Sandage and Tammann (1995) calibration of equation (3.2). Specifically, Fusi Pecci *et al.* (1996) find $M_V(RR) = 0.13(\pm 0.07)[Fe/H] + 0.95(\pm 0.09)$. Further observational programs along these lines are likely to reduce the remaining uncertainty in this calibration. Another useful and independent approach will be to obtain more distances to Milky Way globular clusters using main-sequence fitting and comparing a sample of such distances with those derived from the RR Lyrae method.

The calibration uncertainty propagates into a number of derived quantities discussed in this chapter. Relative values from cluster to cluster tend to be less affected. The uncertainty resulting in the derived integrated luminosities of globular clusters

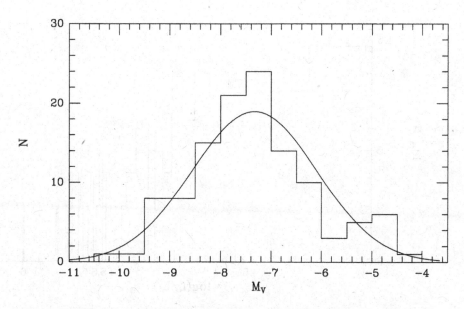

Fig. 3.1. A histogram showing the V-band absolute magnitudes of Milky Way globular clusters. Overlaid is the best-fitting Gaussian to the (unbinned) distribution.

is the most serious, particularly when different globular cluster systems are being compared. We return to this point in later chapters.

3.2 Integrated properties

3.2.1 *The luminosity distribution*

The globular cluster luminosity function is traditionally defined as the relative number of globular clusters per unit magnitude interval. Figure 3.1 shows a histogram of the Milky Way globular cluster luminosity function in V-band. The solid line is a best-fitting Gaussian to the (unbinned) data which excludes the five faintest points. The best-fitting Gaussian to the luminosity function in Figure 3.1 has a mean of $M_V = -7.33$ and a dispersion of 1.23. The 90% confidence limits on these values are $(-7.51, -7.14)$ for the mean and $(1.11, 1.38)$ for the dispersion. These confidence intervals were obtained by bootstrapping, using the ROSTAT statistical package (Beers *et al.* 1990; Bird and Beers 1993).

While the Gaussian is the traditional fit to the distribution of globular cluster luminosities, it is not the only fit, nor is it a particularly good fit. Even when the five faintest Milky Way clusters are excluded, the remaining data are only consistent with a parent Gaussian distribution at the 4% level (Ashman and Bird 1993). This conclusion is based on the skewness of the dataset. Other distributions have been used to fit the data, such as the t_5 distribution (Secker 1992). While this provides a somewhat better fit than the Gaussian, this distribution is, like the Gaussian, symmetric, whereas the skewness indicates that the observed distribution is not. Asymmetric departures from a Gaussian have been studied by Abraham and van den Bergh (1995). These issues are of only minor interest in the context of a general description of the Milky Way

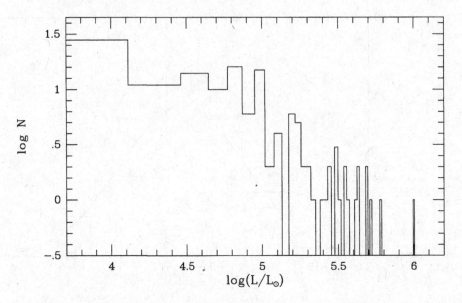

Fig. 3.2. *V*-band luminosities of Milky Way globular clusters. The histogram has bins of constant width in luminosity.

globular cluster system, but, as we show in Chapter 5, they are of importance when comparing the luminosity distributions of different globular cluster systems.

The presence of a peak in this description of the globular cluster luminosity function has influenced ideas concerning globular cluster formation, with the peak being identified with a characteristic globular cluster luminosity (or, more physically, mass). However, it has also been realized that this peak is to some extent a consequence of the logarithmic nature of the magnitude scale. An alternative description of the globular cluster luminosity function can be made in terms of the number of clusters per luminosity interval, $N(L)$ (Surdin 1979; Racine 1980; Richtler 1993; Harris and Pudritz 1994; McLaughlin 1994).

A histogram of the Galactic $N(L)$ distribution is plotted in Figure 3.2. Despite the small numbers, the flattening of the distribution at low luminosities is clearly apparent. The distribution can be fit quite well by two power laws: the best-fitting power law for the high-luminosity portion of the distribution has an exponent of around −1.8 to −2.0, and the best-fitting value for the low-luminosity portion is around −0.2. The uncertainties in the slopes are correlated and are dominated by small number statistics. The transition between the two slopes occurs at a luminosity of about 10^5 L$_\odot$, which is similar to the 'peak' luminosity of the Gaussian fit to the distribution of magnitudes. That both these distributions can fit the available data is not surprising since, over a sufficiently limited range, the Gaussian is similar in form to the logarithm of a power law. As we discuss in Chapter 7, the power-law exponent of the bright end of the luminosity function may be invariant between different galaxies, thereby providing constraints on models of the formation and evolution of globular cluster systems.

3.2.2 Masses and mass-to-light ratios

The mass distribution of Milky Way globular clusters is less amenable to direct observation than the luminosity distribution. Ideally one would like to obtain accurate stellar velocities and surface brightness profiles in order to construct dynamical models and thus masses (see Section 2.6), but suitable velocities are only available for a minority of Milky Way globulars. Mandushev *et al.* (1991) used single-mass King (1966) models to determine masses for 32 Milky Way globulars with measured central velocity dispersions. From this sample they obtained a globular cluster mass–luminosity relationship, which they applied to a further 115 Milky Way globulars with measured luminosities. While this second step requires the assumption that the mass–luminosity relationship found for the subsample of 32 objects applies to all Milky Way globulars, it does allow a mass distribution to be derived. As mentioned in Section 2.6, the use of single-mass models is a simplification which tends to underestimate the globular cluster mass.

With these caveats, Mandushev *et al.* (1991) derived a mean mass for Milky Way globular clusters of 1.9×10^5 $M_\odot$ and a median mass of 8.1×10^4 $M_\odot$. The masses range from 2.6×10^2 $M_\odot$ (AM-4) to 2.4×10^6 $M_\odot$ (ω Cen). The derived mass-to-light ratios vary from 0.66 to 2.9 with a mean value of 1.21. Pryor and Meylan (1993) summarize velocity dispersion results for 56 Galactic globular clusters and, using multi-component King–Michie models (see Section 2.6) derive a mean V-band mass-to-light ratio of 2.3.

Another method of deriving a mass function is to assume a mean mass-to-light ratio for each cluster and calculate a mass from the integrated luminosity. In this case, the mass function obviously has the same form as the luminosity function. The main drawback with this approach is that, while relative masses are probably quite accurate (variations in mass-to-light ratio between clusters are unlikely to be large – see subsection 3.2.4 below), absolute mass values may not be. This is primarily because of uncertainties in the stellar population synthesis models used to transform between luminosity and mass.

3.2.3 Globular cluster radii

There are several ways of characterizing the radii of globular clusters, such as the core radius, r_c, and tidal radius, r_t, described in Chapter 2. However, the core radius is susceptible to change through dynamical processes, whereas the tidal radius is determined by the location within the Galaxy and the orbital characteristics of a given globular cluster. Specifically, the density of a globular cluster cannot be less than the local density due to the surrounding Galaxy, giving a tidal radius of:

$$r_t = R(M_{GC}/3M_{MW})^{1/3}, \tag{3.3}$$

where R is the Galactocentric distance of the cluster, M_{GC} is the globular cluster mass, and M_{MW} is the mass of the Milky Way within R. The expression is general and can be applied to any host galaxy.

A more useful property for studying the *initial* sizes of globular clusters is the half-light radius, $r_{1/2}$, defined in Chapter 2. Numerical simulations suggest that this quantity remains relatively constant over periods as long as ten cluster relaxation

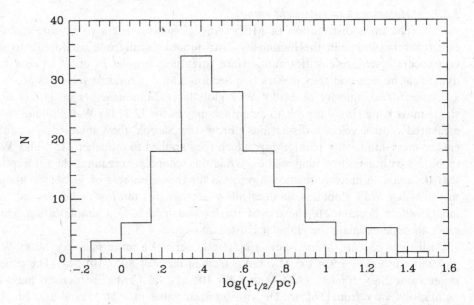

Fig. 3.3. A histogram of the distribution of half-light radii of Milky Way globular clusters.

times (Spitzer and Thuan 1972; Lightman and Shapiro 1978; Murphy *et al.* 1990; see also van den Bergh *et al.* 1991a).

The distribution of $r_{1/2}$ for Milky Way globular clusters is illustrated in Figure 3.3, where we plot this quantity logarithmically. The linear distribution is skew with a tail of clusters with large radii. The distribution has a mean of 4.4 pc and a median of 3.0 pc. The range of radii exhibited by the majority of clusters is quite narrow. Further, an appreciable amount of the dispersion in $r_{1/2}$ is produced by the increase in this quantity with Galactocentric radius (see Section 3.4 below). It is also worth noting that much more mass is concentrated within the half-light radii of globular clusters than other star clusters and associations.

3.2.4   *Correlations and their absence*

One of the most striking aspects of the properties of Milky Way globular clusters is the lack of correlations between them (e.g. van den Bergh *et al.* 1991a). Some of the most detailed statistical analyses of correlations between globular cluster properties have been carried out by Djorgovski and collaborators (e.g. Djorgovski 1993; Djorgovski and Meylan 1994; Djorgovski 1995). An earlier investigation with more limited data was carried out by Brosche and Lentes (1984). The basic technique employed in these studies is that of multivariate statistical analysis (see Murtagh and Heck (1987) for a general description).

Several important results and quantitative confirmations of earlier claims have emerged from these studies. Metallicity does not correlate with any other globular cluster property, with the exception of position within the Galaxy (discussed further in Section 3.4 below) and the stellar mass function slope. Brighter clusters typically

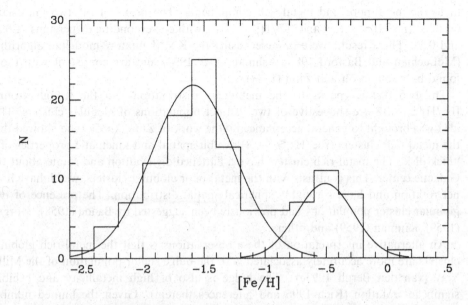

Fig. 3.4. The metallicity distribution of Milky Way globular clusters. Also shown are
the individual Gaussian fits to the two populations determined by mixture-modeling.

have smaller, denser cores but at all luminosities there is considerable scatter in this
relationship. Djorgovski and Meylan (1994) find a deficiency of high-luminosity, low-
concentration clusters and low-luminosity, high-concentration clusters. These authors
attribute the former to initial conditions and the latter to globular cluster destruction
through evaporation (see Section 3.5 below). The two most notable correlations, other
than those depending on position within the Galaxy, involve globular cluster velocity
dispersion which increases with both luminosity and central surface brightness.

Djorgovski (1995) uses some of these results to define a 'Fundamental Plane'
for globular clusters, following an approach applied previously to elliptical galaxies
(e.g. Dressler *et al.* 1987; Djorgovski and Davis 1987). Of particular note is the
scaling relation involving core radius, central velocity dispersion, and central surface
brightness:

$$r_c \propto \sigma^{2.2 \pm 0.15} I_0^{-1.1 \pm 0.1}. \tag{3.4}$$

The form of this relation is close to that expected from the virial theorem:

$$r_c \propto \sigma^2 I_0^{-1} (M/L)^{-1}, \tag{3.5}$$

where (M/L) is the stellar mass-to-light ratio within the cluster. This result suggests
that variations in (M/L) between globular clusters are not large.

3.3   The disk and halo globular cluster systems

In Figure 3.4 we show a histogram of the metallicity distribution of Milky
Way globular clusters. The most striking feature of this distribution is the presence
of two distinct peaks. The overlaid curves in Figure 3.4 represent two Gaussians

fit to the metal-poor and metal-rich components. The peaks of the two Gaussians occur at [Fe/H] = − 1.59 and [Fe/H] = − 0.51 with corresponding dispersions of 0.34 and 0.23. These results were obtained using the KMM mixture-modeling algorithm (McLachlan and Basford 1988; Ashman *et al.* 1994) and are consistent with those found by Armandroff and Zinn (1988).

The two distinct peaks in the metallicity distribution and the trough around [Fe/H] ≃ − 0.8 are suggestive of two distinct populations of globular clusters. This idea was brought to general acceptance by the work of Zinn (1985), who showed that the metal-rich clusters ([Fe/H] > − 0.8) exhibit spatial and kinematic properties of a 'thick disk'. The metal-rich clusters have a flattened distribution and rotate about the Galactic center. This contrasts with the metal-poor globular clusters, which have little net rotation and have a roughly spherical spatial distribution. The presence of two globular cluster populations had previously been suggested by Baade (1958), Morgan (1959), Kinman (1959) and others.

An alternative interpretation of these observations is that the metal-rich globular clusters are more accurately associated with the bulge stellar population of the Milky Way (van den Bergh 1993c). The bulge is also of high metallicity and exhibits significant rotation (Rich 1996 and references therein). Given the limited number of objects, we feel it is difficult to distinguish a bulge population from a thick disk population using only the spatial distribution, although it is worth noting that the metal-rich globular cluster system extends to around 8 kpc from the Galactic center, whereas the bulge is considerably more spatially concentrated. Rich (1996) points out that the rotation of the metal-rich globular cluster system is somewhat greater than that of the bulge, favoring identification of these clusters with the thick disk. Minniti (1995) notes similarities in the stellar populations of the Galactic bulge and metal-rich globular clusters closest to the Galactic center. It is certainly possible that, in this sense, some metal-rich clusters may be related to the bulge population of the Milky Way (see also Zinn 1996).

Perhaps the greatest problem in deciding this issue is the ongoing debate concerning the relationships between the various stellar populations in the Milky Way (e.g. several contributions in Morrison and Sarajedini (1996)). Given the small number of metal-rich globular clusters in the Milky Way, the question of whether some are bulge objects may ultimately be solved indirectly by studies of similar objects around other spiral galaxies.

Another important tool in understanding divisions in the Galactic globular cluster system is the use of proper motion studies to derive space motions of individual clusters (e.g. Rees and Cudworth 1991; Majewski 1994b; Cudworth and Hanson 1994; Zinn 1996). Such studies have confirmed that many metal-rich clusters have space motions expected for members of a thick disk population (e.g. Cudworth and Hanson 1994), although there are also cases where relatively metal-poor clusters appear to have disk-like motions (Rees and Cudworth 1991; Dinescu *et al.* 1996). Further programs of this kind will be extremely important in refining our understanding of the kinematics of the Galactic globular cluster system.

Interestingly, apart from the systemic properties that distinguish the halo and disk systems, the two populations of clusters seem otherwise indistinguishable. For example, the luminosity functions of the two systems show no statistically significant

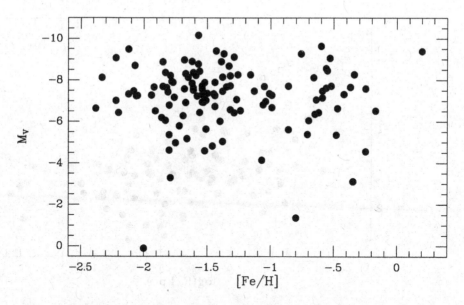

Fig. 3.5. The *V*-band magnitude of Milky Way globular clusters plotted against metallicity.

differences (see Armandroff 1993, and references therein). In Figure 3.5 we plot globular cluster magnitude against metallicity. The lack of a discernable trend between these two quantities provides important constraints on the globular cluster formation models discussed in Chapter 7.

Irrespective of the issues discussed above, the division between these clusters and the metal-poor halo clusters is extremely important when analyzing the properties of the Milky Way globular cluster system. Lumping all globular clusters together tends to obscure important apsects of each subsystem.

3.4   Trends with Galactocentric distance

3.4.1   *Metallicity gradients*

The separation of the Milky Way globulars into disk and halo systems is essential when discussing the question of metallicity gradients. Without this distinction, the overall system is found to show a negative metallicity gradient, primarily driven by high-metallicity clusters within about 8 kpc of the Galactic center (e.g. Searle and Zinn 1978; Zinn 1985). This trend is illustrated in Figure 3.6. If the disk clusters are removed, however, there is no evidence for a significant metallicity gradient in the halo clusters beyond 7 kpc (Zinn 1985). There *is* some evidence that the inner halo clusters ($R < 7$ kpc) have a higher mean metallicity than those beyond 7 kpc, so that in this limited sense there may be a slight negative metallicity gradient in the halo globular cluster system (Zinn 1985; Armandroff *et al.* 1992). Even this possibility is unclear, however, since the space velocity results discussed in the previous section suggest that there may be a metal-poor tail to the disk globular cluster system. Such clusters are still more metal-rich than the halo mean and would tend to increase

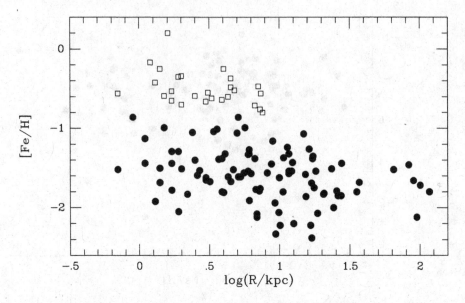

Fig. 3.6. Metallicity of Milky Way globular clusters plotted against Galactocentric distance. Open squares represent disk clusters, filled circles represent halo clusters.

estimates of the mean metallicity of the inner halo clusters above the value for genuine members of this population. With more space velocities, it will be feasible to make a disk–halo division on the basis of kinematics rather than metallicity, providing a definitive answer to the question of a metallicity gradient in the halo clusters.

There is also tentative evidence for metallicity gradients within the disk system, both with R and with distance from the Galactic plane, Z (Zinn 1985; Armandroff 1989, 1993). The trend with R does not appear to be statistically significant, but Armandroff (1993) finds the gradient in Z to be significant at the 2.7σ level.

3.4.2   *Stellar radial distributions*

Variations between globular cluster properties and position within the Galaxy may arise through evolutionary processes. Alternatively, they may reflect the properties of globular clusters at the time of their formation. Clearly it is important to be able to distinguish between these two possibilities. The globular cluster half-light radius is of considerable interest in this regard, since it is believed to be largely unaffected by evolutionary processes. It is therefore of significance that $r_{1/2}$ is found to increase roughly as $R^{1/2}$ (van den Bergh *et al.* 1991a). This trend is illustrated in Figure 3.7. Note that because there is no observed correlation between globular cluster mass (luminosity) with half-light radius, this trend implies that globular cluster stellar density decreases with R.

The absence of diffuse clusters close to the Galactic center may be a consequence of dynamical disruption. It is difficult for such objects to survive the various dynamical destruction processes (see Section 3.5), although, clearly, one cannot rule out the

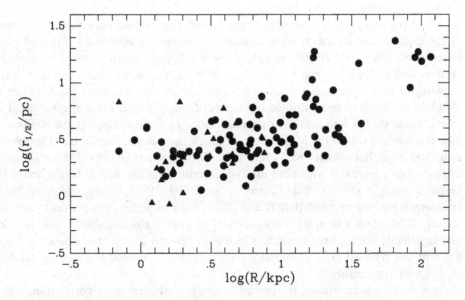

Fig. 3.7. The half-light radii of Milky Way globular clusters plotted against Galacto-centric distance. Halo clusters are represented by circles, disk clusters are represented by triangles.

possibility that such objects never formed. Less ambiguous is the absence of compact clusters beyond about 8 kpc from the Galactic center. Such objects would be extremely resilient, comfortably surviving to the present epoch, so it seems likely that such clusters simply did not form. This also suggests that much of the increase of $r_{1/2}$ with R was probably set at the epoch of globular cluster formation (see van den Bergh *et al.* 1991a). Another implication of these results is that some of the dispersion in the size distribution of Milky Way globulars is being driven by the increase of $r_{1/2}$ with R. Thus at a given Galactocentric distance, the dispersion in $r_{1/2}$ is even less than that for the whole system quoted in Section 3.2 above. An additional finding by van den Bergh *et al.* (1991a) is that both disk and halo globular clusters follow the same $r_{1/2}$–R relation. This is illustrated in Figure 3.7, where the two populations are plotted with different symbols.

Globular clusters with evidence for collapsed cores (Section 2.6) account for about one-fifth of all Milky Way globulars (Djorgovski and King 1986; Lugger *et al.* 1987; Hut and Djorgovski 1992). Such clusters show a tendency to be found closer to the Galactic center than would be expected if they constituted a random sample. Djorgovski and Meylan (1994) have also found a tendency for King-model clusters close to the disk plane or Galactic center to be more concentrated than those at larger Galactocentric distances. Both effects are expected on dynamical grounds (see Section 3.5). Clusters closer to the Galactic center experience more dramatic dynamical effects produced by the gravitational field of the Galaxy, the most important being disk-shocking. This accelerates the dynamical evolution of clusters, including the process of core collapse.

3.4.3   *The stellar mass function*

Several attempts have been made to find correlations between the exponent of the globular cluster stellar mass function, x (defined in equation (2.5)), and other parameters. This field is still hampered by a relatively small number of globular clusters with well-determined values of x, but interesting results have emerged. Early suggestions that x depends primarily on globular cluster metallicity (McClure *et al.* 1986) do not seem to be confirmed by studies of larger datasets (e.g. Capaccioli *et al.* 1991). These studies find that the dominant factor in determining x is the position of globular clusters within the Milky Way, although there is also a mild trend for metal-rich clusters to have flatter mass functions (Djorgovski *et al.* 1993). Specifically, it appears that x increases with both the Galactocentric radius, R, and height above the Galactic plane, Z (Piotto 1991; Capaccioli *et al.* 1991, 1993; Djorgovski *et al.* 1993). It is worth bearing in mind that R and Z are not completely independent variables (clearly $Z \leq R$, for example). Djorgovski *et al.* (1993) also concluded that position and metallicity are the *only* significant factors in determining the mass function slope. Current and planned *HST* observations are expected to expand the current database in this area appreciably.

As is usual in such studies, the central question is whether these correlations reflect different formation histories or evolutionary processes. For the correlations between x and position, an excellent evolutionary candidate exists. As discussed in Section 2.6, mass segregation is expected to be important in globular clusters, with lower-mass stars occupying the outer regions of globular clusters. These stars are more likely to be removed from the cluster through dynamical processes such as disk-shocking (see Section 3.5 below). Thus clusters that experience more passages through the Galactic disk are expected to lose more low-mass stars and thus have flatter mass function slopes (smaller values of x). Since clusters at smaller values of R and Z will, on average, experience more encounters with the Galactic disk, this dynamical effect reproduces the observed trend between x and position qualitatively. Independent evidence that this effect is influencing x comes from the flat (low x) luminosity function of NGC 6397 discussed in Section 2.4.

Stiavelli *et al.* (1991, 1992) and Capaccioli *et al.* (1993) have studied the problem using N-body simulations, assuming that initially all globular mass function slopes are the same. Their results show reasonable agreement with the observed trends. One oddity discussed by Djorgovski *et al.* (1993) is that the empirical correlations show *less* scatter than predicted by the dynamical studies. These authors speculate that this may reflect a correlation between mass function slope and the ability of a globular cluster to survive dynamical disruption. Such a correlation is expected, based on analyses of globular cluster destruction mechanisms (see Section 3.5 below).

In contrast to correlations with position, a dependence of x on metallicity seems more likely to have a primordial origin. There are two possibilities. The most straightforward explanation is that the mass function slope is determined by the metallicity of the gas out of which globular clusters form. Alternatively, if the observed metallicity of globular clusters is produced by self-enrichment, then one expects a correlation between x and metallicity because flatter mass functions produce more high-mass, metal-producing stars. Indeed, the sense of the correlation is the same as the empirical trend. However, there are two drawbacks to the self-enrichment

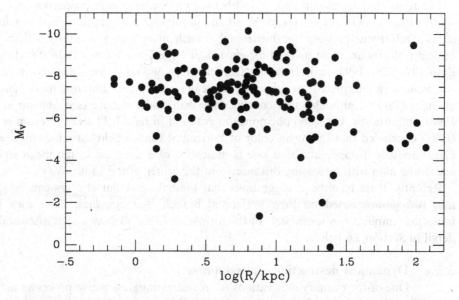

Fig. 3.8. *V*-band magnitude of Milky Way globular clusters plotted against Galacto-centric distance.

scenario. First, it is generally accepted that the disk globular clusters underwent at least some pre-enrichment. Since it is primarily these metal-rich clusters that appear to be driving the correlation between x and metallicity, the idea of self-enrichment being responsible for the trend is difficult to sustain. Further, many self-enrichment models require that the observed stellar population in globular clusters is *not* the same as the population responsible for metal production (see Chapter 7 for a full discussion). A globular cluster self-enrichment formation model that accounts for the correlation between x and metallicity was proposed by Smith and McClure (1987).

3.4.4 Other radial trends

A dependence of globular cluster mass on Galactic radius is predicted by many models of the formation and dynamical evolution of the Galactic globular cluster population. However, little or no effect is observed. Figure 3.8 shows a plot of M_V against R. There is a slight decrease in the typical M_V at very large R, driven by the remote and faint globular clusters beyond the Magellanic Clouds, but no overall trend is apparent. A robust regression on the unbinned data finds no statistically significant evidence for a correlation, if the objects beyond 30 kpc are removed from the dataset. There is evidence for a decrease in M_V with estimated perigalactic distance (van den Bergh 1995b), but this is driven by a handful of faint clusters. This may indicate a dichotomy between the bulk of Milky Way globular clusters and the small number of faint, diffuse clusters at large Galactocentric distances. Interestingly, these faint clusters also tend to have the largest second parameter effects in their horizontal branch morphologies (van den Bergh 1995b).

The remaining significant correlation between globular cluster properties and position within the Galaxy relates to the second parameter problem described in Chapter 2. Differences between the horizontal branch morphology of clusters with the same metallicity are most marked for clusters at large Galactocetric radii (Searle and Zinn 1978; Lee 1993, and references therein). This 'global' second parameter problem is usually interpreted as indicating a greater age spread amongst more remote globular clusters, although this conclusion is still under debate (see Stetson *et al.* 1996). Importantly, the same phemonenon is found in halo field stars. Preston *et al.* (1991) concluded that the mean color of horizontal branch field stars increases with Galactocentric distance and that this is indicative of a decrease in the mean stellar age in the halo with increasing distance from the center of the Milky Way.

Recently, there have been suggestions that the halo globular clusters can be split into two groups based on their horizontal branch characteristics. This idea has important implications for models of the formation of the Galaxy and is discussed in detail in Section 3.6 below.

3.5   Dynamical destruction and evolution

One of the primary motivations for investigating destructive processes acting on Milky Way globular clusters is to establish the properties of the original Milky Way globular cluster system. In particular, the initial total number and distribution of globular clusters is of great interest. A specific issue that drives this interest is the possibility that the field stars of the Galactic spheroid represent the disrupted remnants of a much larger globular cluster population. In this section we will address this possibility from a dynamical viewpoint. Other constraints from chemical properties and abundance patterns are combined with these dynamical constraints in Chapter 7.

Globular clusters in the Milky Way and other galaxies are subject to various dynamical processes, the most dramatic example being the destruction of individual globular clusters. A full treatment of this topic and literature overview is provided by Spitzer (1987), Elson *et al.* (1987), Djorgovski and Meylan (1993) and Meylan and Heggie (1996). Most dynamical studies have concentrated on long-term destructive dynamical processes. However, it is worth noting that the most precarious stage in the evolution of a globular cluster may be soon after it has formed, when short-term stellar evolutionary processes can lead to disruption. Since this possibility is intimately linked with the formation process itself, we defer a discussion of this topic to Chapter 7.

3.5.1   Analytical studies

The three dominant dynamical processes that are usually considered in the context of globular cluster systems are evaporation, disk-shocking and dynamical friction. Two-body encounters between stars in a globular cluster tend to drive the velocity distribution towards a Maxwellian. Stars in the high-velocity tail of this distribution acquire enough energy to escape from the cluster altogether. This is the process of evaporation (see Spitzer and Thuan 1972; Spitzer 1975, 1987). Isolated clusters experience evaporation, but the presence of an external gravitational field (due to the host galaxy) tends to make the process more efficient. The evaporation process is also accelerated if stars in a globular cluster achieve energy equipartition, so that all stars have the same kinetic energy. Since the stellar population of a globular cluster is comprised of a range of stellar masses, this means that the more massive stars slow down and sink towards the central regions of the cluster, whereas the less massive stars have higher velocities than the mean and tend to occupy the outer cluster regions. This process of mass segregation therefore leads to the preferential loss of low-mass stars through evaporation. In fact, stars in the outer parts of globulars are the most susceptible to the other destructive dynamical processes described below. As an aside, this may have a bearing on the mass-to-light ratios of globular clusters which are somewhat lower than expected for old stellar populations (Section 2.6), since the low-mass stars that are most likely to escape have the highest mass-to-light ratios (see Ostriker et al. 1972).

Disk-shocking (Ostriker et al. 1972) occurs when a globular cluster crosses the high-density disk of the Milky Way. (Shocks due to the Milky Way bulge have a similar effect.) The result of the gravitational shock produced by the disk is to transfer energy to the stars within the globular cluster. The increase in stellar velocities leads to stars being lost from the cluster. Recent theoretical work has refined the analysis of disk-shocking and indicates that the destructive effects of this process may be considerably more important than indicated by equation (3.7) below (Weinberg 1994a,b; Kundić and Ostriker 1995).

Globular clusters are also subject to dynamical friction (see Binney and Tremaine 1987). In principle, this process causes globular clusters to spiral in towards the Galactic center. The likely end-point for such clusters is disruption before they reach the Galaxy center, due to the high stellar densities and resulting strong tidal field of the Galaxy at small radii.

Early attempts to assess the importance of these effects used approximate timescales over which the dynamical processes are expected to disrupt globular clusters. Because the effects depend on the surrounding gravitational field, the timescales are a function of Galactocentric radius. A useful form of these timescales has been provided by Okazaki and Tosa (1995), based on earlier analytical work. Assuming that a typical cluster is comprised of stars with a mean mass of 0.5 $M_\odot$, the evaporation timescale can be expressed as:

$$t_{evap} \simeq 8 \times 10^9 \left(\frac{M_c}{10^5 \, M_\odot} \right)^{1/2} \left(\frac{r_t/r_{1/2}}{10} \right) \left(\frac{r_{1/2}}{1 \, pc} \right)^{3/2} yr, \qquad (3.6)$$

where M_c is the total mass of the cluster and r_t and $r_{1/2}$ are the tidal and half-light radii defined previously. (The normalization of equation (3.6) differs from that given by Okazaki and Tosa (1995) who assumed a mean stellar mass of 1 $M_\odot$.)

The timescale for disk-shocking is roughly:

$$t_{sh} \simeq 5 \times 10^{12} \left(\frac{v_c}{220 \, kms^{-1}} \right) \left(\frac{M_c}{10^5 \, M_\odot} \right) \left(\frac{R}{1 \, kpc} \right) \left[\exp \left(-\frac{R - R_\odot}{h_R} \right) \right]^{-2} \left(\frac{r_{1/2}}{1 \, pc} \right)^{-3} yr, \qquad (3.7)$$

where R is the Galactocentric distance of the cluster and h_R is the scale-height of the Milky Way disk. This equation assumes that the cluster follows a circular orbit with a rotation velocity v_c.

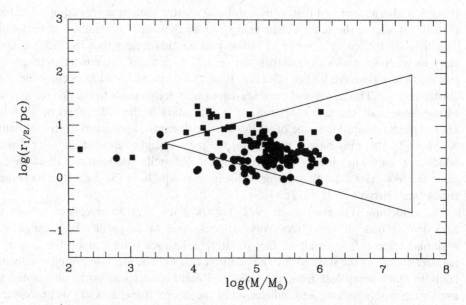

Fig. 3.9. The survival triangle of Milky Way globular clusters, calculated for a Galactocentric distance of $R = 8.5$ kpc. The three sides of the triangle are determined by: disk-shocking (upper), evaporation (lower) and dynamical friction (vertical line). Filled squares represent clusters with $R > 8.5$ kpc, whereas filled circles represent clusters with $R < 8.5$ kpc.

Finally, the dynamical friction timescale can be written:

$$t_{fric} \simeq 6 \times 10^{10} \left(\frac{R}{1 \text{ kpc}} \right)^2 \left(\frac{v_c}{220 \text{ kms}^{-1}} \right)^2 \left(\frac{\ln \Lambda}{10} \right)^{-1} \left(\frac{M_c}{10^5 \text{ M}_\odot} \right)^{-1} \text{ yr}, \qquad (3.8)$$

where $\ln \Lambda$ is the Coulomb logarithm, normalized to a value appropriate for the problem under study (see Binney and Tremaine 1987).

The basic interpretation of all these timescales is that they represent the characteristic elapsed time before a globular cluster is destroyed. Several results follow immediately from the equations. Low-mass clusters are particularly susceptible to evaporation, whereas globular clusters close to the Galactic center are vulnerable to disk-shocking and dynamical friction. In fact, for clusters at very small R, the impulse approximation upon which equation (3.7) rests breaks down (Spitzer 1958). This is because the short time between disk passages means the shock is no longer impulsive. The result is that, at small R, disk-shocking is not quite as efficient as the above expression suggests. The strong dependence on R of the dynamical friction timescale suggests that massive clusters initially within about 1 kpc of the Galactic center cannot survive to the present epoch. Equation (3.8) further suggests that, much beyond this radius, the effects of dynamical friction are negligible.

One of the central ideas in globular cluster destruction was advanced by Fall and Rees (1977). These authors noted that the dominant long-term disruption processes acting on Milky Way globular clusters define a 'survival triangle' in the mass–radius plane of globular clusters (see Figure 3.9). The triangle is formed by setting the three timescales given in equations (3.6)–(3.8) above equal to the Hubble time (comparable to the age of the Milky Way), which we assume to be 1.5×10^{10} yr. For the sake of illustration, the timescales are calculated at the solar Galactocentric distance of $R = 8.5$ kpc. Fall and Rees (1977) suggested that globular clusters initially had a wide range of masses and radii, but that only those within the survival triangle would remain intact to the present epoch. Figure 3.9 shows that observed globular clusters do, for the most part, fall within the triangle, although it appears that the upper mass of globulars is *not* set by the dynamical friction condition, which is only relevant for objects with a considerably greater mass. Fall and Rees (1977) recognized this problem and suggested that the initial population followed a mean mass–radius trend that passed through the triangle. (Such a trend is actually expected if globular clusters are formed from a primordial spectrum of density fluctuations, as in the models discussed in Section 7.2.) In this way, only two sides of the triangle are required to reduce an initial population with a broad distribution of masses and radii to the observed population.

The different symbols in Figure 3.9 represent globular clusters with $R < 8.5$ kpc (circles) and $R > 8.5$ kpc (squares). It is interesting to note that all the clusters falling above the disk-shocking limit (upper line in Figure 3.9) have $R > 8.5$ kpc. These clusters are less susceptible to disruption through shocks, so their survival is not surprising. (To put it another way, the survival triangle for large R has a disk-shocking limit above that shown in Figure 3.9.) Another interesting point concerns those clusters with masses between about 10^4 M$_\odot$ and 10^5 M$_\odot$ below the lower line representing evaporation. More than half of these clusters are core-collapsed or have high concentrations, so that they are more resilient than typical clusters to evaporation (cf. the dependence of t_{evap} on $r_t/r_{1/2}$ in equation (3.6)).

There are a couple of problems with the survival triangle idea (e.g. Gunn 1980; Caputo and Castellani 1984; Fall and Rees 1985). First, the universality of the globular cluster luminosity function (Chapter 5) suggests that whatever process is responsible for restricting the distribution of globular cluster mass and radius should operate equally effectively in both spiral and elliptical galaxies. Since elliptical galaxies do not have disks, it is difficult to see how disk-shocking could be important in such systems, although it is plausible that passage through the bulge of an elliptical could have a similar disruptive effect on globular clusters. The fact that the mass of globulars in the Milky Way is roughly independent of Galactocentric radius poses a related problem, since clusters at large radii are less likely to interact with the Galactic disk (see Caputo and Castellani 1984). This problem may be mitigated by a preponderance of radial globular cluster orbits.

3.5.2 Numerical studies

While the above analytical results provide important insights into the dynamical destruction of globular clusters and evolution of the Galactic globular cluster system, they are somewhat limited by the approximations upon which they are based. Further, it is difficult to use these expressions to study the evolution of a system of globular clusters with a range of properties and orbital characteristics. This has led to more emphasis being placed on numerical studies of the problem (e.g. Chernoff *et al.* 1986; Chernoff and Shapiro 1987; Aguilar *et al.* 1988; Chernoff and Weinberg

1990; Oh *et al.* 1992; Oh and Lin 1992; Capaciolli *et al.* 1993; Weinberg 1994c; Okazaki and Tosa 1995; Gnedin and Ostriker 1996). Here we focus primarily on the key results to emerge from these studies. Details of the numerical methods used can be found in the original sources.

The internal dynamical timescales of globular clusters are generally much shorter than their orbital timescales, so that complete *N*-body simulations of the Milky Way globular cluster system are extremely difficult. In practice, numerical studies have employed various simplifying assumptions and approximations and/or have concentrated on specific dynamical processes. As more of these studies have been conducted, the modeling has become more sophisticated and a greater area of parameter space has been explored. Clusters with a range of masses, concentrations, stellar mass functions, Galactocentric positions, and orbits have been subjected to the various dynamical effects of interest.

Tidal heating of globular clusters has received particular attention. Chernoff *et al.* (1986) examined the relative importance of tidal heating due to disk-shocking and globular cluster interactions with Giant Molecular Clouds. The latter effect has been studied analytically by Wielen (1985, 1988). Chernoff *et al.* (1986) found that disk-shocking is the more effective in terms of its disruptive influence on globular clusters. At small Galactocentric radii (within about 3 kpc), disk-shocking drives most clusters towards core collapse. At slightly greater distances, typical globular clusters either suffer core collapse or are disrupted, with clusters that are initially less concentrated being more susceptible to disruption. Mass loss produced by disk-shocking is greater than that associated with evaporation caused by internal relaxation for clusters within about 8 kpc of the Galaxy center. Weinberg (1994a,b) has refined the analysis of tidal heating. His numerical study (Weinberg 1994c) concluded that disk-shocking dominates globular cluster evolution for objects within 8 kpc of the Galactic center.

The influence of the stellar mass function of globular clusters on their ultimate fate was addressed using approximate methods by Chernoff and Shapiro (1987) and more comprehensively by Chernoff and Weinberg (1990). For a stellar mass function of the form given in equation (2.6), it was found that stellar mass loss dominates the evolution for the first 5 Gyr for clusters with $x < 2.0$. For a Salpeter mass function ($x = 1.35$) only concentrated clusters ($W_0 > 4.5$, where W_0 is the concentration parameter defined in Section 2.6) survive disruption due to stellar mass loss. However, if a cluster does survive for more than 5 Gyr, further stellar mass loss has a negligible effect on the cluster and its location and total mass determine its fate.

These studies confirmed the expectation from analytical work that evolution is more rapid in clusters with a range of stellar masses. (This is probably due mainly to energy equipartition producing mass segregation, which accelerates both core collapse and the loss of low-mass stars from the cluster.) The tendency of lower-mass clusters to be more easily disrupted was also confirmed, as was the fact that for reasonable globular cluster masses dynamical friction has negligible effect. Clusters at large Galactocentric distances are mostly unaffected by these dynamical processes, unless they are on plunging orbits that take them through the Galactic disk and bulge. Even in this case, the relatively long orbital periods and resulting infrequency of disk or bulge crossings makes the effect small for many clusters. One other interesting result to emerge is that core collapse can delay evaporation by increasing the concentration of the cluster.

Lee and Goodman (1995) have studied the evolution of globular clusters that have already experienced core collapse. Even at this stage, clusters are far from safe. Stellar dynamical processes in the high-density collapsed core can reduce the overall binding energy of the cluster, resulting in stellar evaporation. Lee and Goodman (1995) find that this process alone may have destroyed a greater number of globular clusters than the current total population.

The key quantity in determining the original number of Milky Way globular clusters is the globular cluster destruction rate. One of the first numerical studies aimed at estimating this rate was carried out by Aguilar *et al.* (1988). These authors developed a kinematic model of the *current* globular cluster system based on observed positions and radial velocities of globular clusters along with randomized values of the other two cluster velocity components. Their findings included a quantitative confirmation that clusters on highly eccentric orbits are susceptible to disruption from disk- and bulge-shocking. The central result found by Aguilar *et al.* (1988) was a current globular cluster destruction rate of 0.5 clusters/Gyr.

Using a similar philosophy, but with a more sophisticated treatment and the inclusion of the Kundić and Ostriker (1995) disk-shocking term, Gnedin and Ostriker (1996) obtain a present destruction rate around an order of magnitude higher than Aguilar *et al.* (1988). Quantitatively, they find that more than half the current Milky Way globular clusters will be disrupted over the next Hubble time. This number is consistent with the destruction rate of 5 ± 3 clusters/Gyr found by Hut and Djorgovski (1992) based on observed relaxation times of globular clusters and analytical disruption timescale estimates.

A value of the present destruction rate does not provide a direct answer to the fundamental question of the original number of Milky Way globular clusters. Gnedin and Ostriker (1996) take an important step in this direction by using their results to estimate the time evolution of the destruction rate. There are many uncertainties in this procedure, but the results of Gnedin and Ostriker (1996) suggest that the original number of clusters may have been at least an order of magnitude greater than the current population. Further, these dynamical studies indicate that it is primarily low-mass clusters that are most susceptible to destruction.

Another area of interest concerning the original globular cluster system is the possibility that the globular cluster mass function has evolved significantly over time. This possibility seems fairly natural, given the result that low-mass clusters are more easily disrupted. Okazaki and Tosa (1995) use a similar technique to Chernoff and Weinberg (1987) in an attempt to assess how an initial power-law mass function of globular clusters evolves due to dynamical evolution. (Part of the motivation is the evidence that systems of candidate young globular clusters have a luminosity function similar to a power law – see Chapter 6.) Okazaki and Tosa (1995) claim the observed log-normal form of the present-day mass function can be reproduced from an initial power-law form through preferential destruction of low-mass clusters. However, these authors only considered evolution of clusters within 13 kpc of the Galactic center. As already mentioned, clusters beyond this radius are largely unaffected by dynamical processes. Since clusters with $R > 13$ kpc have a mean mass that is essentially indistinguishable from inner clusters (see Figure 3.8), it seems unlikely that dynamical evolution alone is responsible for the currently observed mass function. (Further, the marginal evidence for a decrease in globular cluster mass for some of the most remote globular clusters appears to be driven by an absence of high-mass clusters rather than dynamical destruction of low-mass clusters.) We return to this point in Chapters 6 and 7, where observations of extragalactic globular cluster systems and ideas from globular cluster formation models are used to provide further constraints on the evolution of the globular cluster mass function.

Other work in this area includes the study by Oh and Lin (1992) which paid particular attention to the effects of tidal heating and evaporation on the orbits of stars within globular clusters. These considerations may modify disruption timescales and also provide potential observational tests of the importance of tidal heating.

Most dynamical studies of globular clusters have been applied to the Milky Way system. There are, however, important exceptions which have provided clues to the evolution of globular clusters in elliptical galaxies. Such studies also have important implications for the interpretation of observed differences between globular cluster systems. We discuss this work and its connection to the material of this section in Chapter 6.

3.6 Globular clusters and the formation of the Galactic halo

In this section we examine the connection between the halo globular clusters and the field stars of the Milky Way halo. This connection is important in under-standing the formation of the visible spheroid of the Milky Way. Similarities and differences between spheroid stars and globular clusters also shed light on the process of globular cluster formation, discussed further in Chapter 7.

3.6.1 Comparing halo stars and globular clusters

The Milky Way halo globular clusters make up about 1% of the stars in the visible halo. (We regard the central Galactic bulge as distinct from this halo.) There are notable similarities and differences between the properties of the field stars and globular clusters. The spatial distributions of halo stars and globular clusters in the Milky Way appear to be indistinguishable (Zinn 1990). Both populations are dynamically hot, with little net rotation, and contain a significant proportion of objects on retrograde orbits. We discuss comparisons between halo field stars and globular clusters around other galaxies in Chapters 4 and 5.

The peak of the metallicity distribution of stars and globular clusters in the Milky Way halo is also similar. However, the metallicity *distributions* of the two popula-tions exhibit differences, with some field stars having much lower metallicities than the most metal-poor globular clusters. Qualitatively, such a difference is expected even if the parent metallicity distributions are identical, simply because the field stars are far more numerous than the globular clusters. However, analyses by Laird *et al.* (1993) and Carney *et al.* (1996) indicate that there *is* a statistically signifi-cant difference between the two metallicity distributions that cannot be explained by the lower numbers of globular clusters. Relative to the field stars, there is a deficiency of metal-poor globular clusters (see Figure 3.10). There are also abundance differences in elements such as C, N and O between halo field stars and globu-

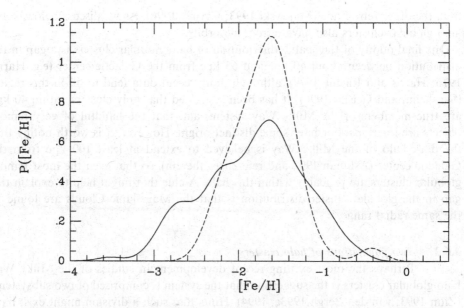

Fig. 3.10. A comparison of the metallicity distribution of halo globular clusters (solid line) and halo field stars (dashed line). (From data supplied by J. Laird.)

lar clusters (see Section 2.4). Field stars are almost all CN-weak, whereas many globular clusters have bimodal CN distributions with some stars having high CN abundances.

Comparisons of the kinematics and abundance gradients of halo stars and globular clusters are complicated by problems of obtaining a 'pure' sample of halo stars (see Majewski 1993; also Norris and Ryan 1989). This continues to be a controversial area of research, but the bulk of the evidence appears to favor no significant metallicity gradient in Galactic halo stars (e.g. Saha 1985; Ratnatunga and Freeman 1989; Suntzeff *et al.* 1991; Croswell *et al.* 1991; Majewski 1992). Studies which do suggest a gradient (e.g. Sandage and Fouts 1987) tend to include stars at smaller distances above the Galactic disk, so it is possible that a metallicity gradient in halo stars is present relatively close to the Galactic plane. However, beyond about 3 kpc from the Galactic plane, the halo stars, like the halo globular clusters, appear to have no metallicity gradient.

Determining the kinematics of halo stars and globular clusters provides an observational challenge and results obtained by different groups often differ significantly (Majewski 1993). Perhaps the only safe conclusion is that, like the halo globular clusters, the rotation of the stellar halo is not large. Many studies report a slight prograde rotation of around 30 km/s (e.g. Norris 1986; Freeman 1987; Gilmore *et al.* 1989). This compares to a small, prograde rotation in the halo globular clusters of about 40 ± 30 km/s (Frenk and White 1980; Zinn 1991). However, more recent surveys have found *retrograde* rotation in the halo field stars (Reid 1990; Allen *et al.* 1991; Majewski 1992; Schuster *et al.* 1993). A possible resolution of this apparent discrepancy is that there are two populations of halo stars, one with prograde rotation, the other retrograde (Majewski 1993; Carney 1996). As we discuss below, some halo globular clusters also have retrograde orbits.

One final oddity of the spatial distribution of halo globular clusters is a gap in the distribution between about 40 kpc and 65 kpc from the Galactic center (e.g. Harris 1976; Harris and Racine 1979), although more recent data tend to fill in this region (Sarajedini and Geisler 1996). It has been suggested that only clusters within 40 kpc are true members of the Milky Way system, and that the handful of very remote objects are interlopers or have some distinct origin. However, it is worth noting that the dark halo of the Milky Way is believed to extend at least 100 kpc from the Galactic center (Ashman 1992 and references therein), so that even the most remote globular clusters are probably within this halo. A clue that might help to explain this gap in the globular cluster distribution is that the Magellanic Clouds are found in the same radial range.

3.6.2   *Two populations of halo clusters*

Perhaps the most exciting recent development in studies of the Milky Way halo globular clusters is the suggestion that the system is comprised of two subsystems (Zinn 1993; van den Bergh 1993a, 1994). Hints that such a division might exist have been around for some time. To understand the early evidence favoring two populations of halo globular clusters and to appreciate the importance of the more recent results, a brief discussion of galaxy formation models is warranted.

There are two extreme paradigms of galaxy formation that are frequently used as benchmarks for more sophisticated treatments. The first view is that a galaxy forms out of a monolithic gas cloud which contracts as it cools (dissipative collapse). Star formation during the collapse enriches the surviving gas, leading to a metallicity gradient in the stellar population. Angular momentum conservation causes the gas to 'spin-up' and become more flattened as it collapses.* The alternative extreme, often associated with Searle and Zinn (1978) is that galaxies (specifically the Milky Way halo) are assembled from a large number of gaseous subunits in a rather chaotic manner. Since the subunits are supposed to evolve relatively independently, no metallicity gradient is expected.

Searle and Zinn (1978) found no metallicity gradient in the globular cluster system if clusters inside 8 kpc were ignored. Rodgers and Paltoglou (1984) noted that a group of globular clusters with metallicities between −1.3 and −1.7 all appeared to have retrograde orbits. Both findings are naturally explained in a chaotic collapse model, but are extremely difficult to reproduce through monolithic dissipative collapse. On the other hand, Hartwick (1987) found that globular clusters in the inner halo have a somewhat flattened spatial distribution around the Galactic disk (even after obvious disk clusters have been removed), whereas more remote globulars follow a roughly spherical distribution. Further, Armandroff (1989, 1993) found weak metallicity gradients with Galactocentric distance and height above the Galactic plane for globular clusters in the inner halo (see Section 3.4 above). These two results support

---

* Monolithic dissipative collapse is often associated in the literature with the Galaxy formation model of Eggen, Lynden-Bell and Sandage (1962). In fact, these authors did not specifically argue for a monolithic collapse and their model does *not* predict a metallicity gradient in the Milky Way halo. The origin of the misconceptions surrounding this picture are described by Majewski (1993).

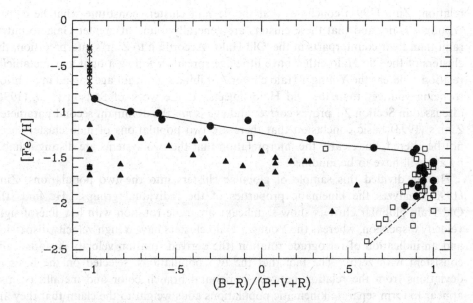

Fig. 3.11. Metallicity against horizontal branch type for Milky Way globular clusters. The different symbols represent disk clusters (crosses), 'Younger Halo' clusters (filled triangles), 'Old Halo' with $R < 6$ kpc (filled circles) and 'Old Halo' with $R > 6$ kpc (open squares). The solid line is the mean trend through Old Halo clusters with $R < 6$ kpc extended through the disk clusters. The dashed line follows Old Halo clusters with $R > 6$ kpc and the bluest horizontal branches. (From data supplied by R. Zinn.)

a dissipative collapse picture. Finally, Searle and Zinn (1978) presented evidence that globular clusters in the inner regions of the Galaxy follow a tight relationship between horizontal branch morphology and metallicity, whereas clusters at larger Galactocentric radii exhibit a considerable scatter between these two quantities. This finding was supported by the work of Lee *et al.* (1994).

This body of circumstantial evidence pointed towards the possibility of two populations of halo globular clusters. More recent studies have put the possibility on a firmer footing. Zinn (1993) has analyzed the problem by dividing the halo globular clusters into two groups, based on whether or not a cluster follows the relationship between horizontal branch and metallicity of the inner halo clusters. Using the horizontal branch parameter $C \equiv (B - R)/(B + V + R)$ described in Section 2.1 above, Zinn (1993) found a tight relationship between C and [Fe/H] for 'inner' globular clusters with $R < 6$ kpc. When 87 globular clusters with sufficiently detailed color–magnitude diagrams are plotted in this plane, there is a fairly clear division between clusters that follow the mean trend between horizontal branch type and metallicity and those that deviate from this trend (see Figure 3.11). All these deviant clusters have metallicities that are lower than the mean value at their horizontal branch color or, equivalently, have horizontal branches which are too red for their metallicity.

This result suggests the possibility that more remote clusters are younger than the inner clusters of the Milky Way, since it is the latter that follow the color–metallicity relation. Zinn (1993) concludes that the deviant clusters constitute what he calls a 'Younger Halo' and that these clusters are generally found at greater Galactocentric radii than their counterparts in the 'Old Halo'. According to Zinn's interpretation, the clusters of the Old Halo either have little age spread or follow a tight age–metallicity relation, whereas the Younger Halo clusters exhibit a significant age spread in addition to being younger than the Old Halo objects. If the work of Stetson *et al.* (1996) discussed in Section 2.1 proves correct and age is *not* the dominant second parameter, Zinn's (1993) basic conclusion that there are two populations of halo clusters need not be altered, but clearly the interpretation that the two systems are distinguishable by age will have to be modified.

Having divided this sample of globular clusters into the two populations, Zinn (1993) analyzes the kinematic properties of the individual groups. He finds the Old Halo globular clusters show significant prograde rotation with low line-of-sight velocity dispersion, whereas the Younger Halo clusters have a high velocity dispersion and an indication of retrograde rotation (the derived rotation velocity is statistically consistent with zero). The fact that the two populations, selected on the basis of deviations from the relation between horizontal branch color and metallicity, also appear to form separate kinematic populations adds weight to the claim that they are genuinely distinct. Further, Zinn (1993) notes that the Younger Halo clusters show no evidence of a metallicity gradient with Galactocentric radius (as is the case for the entire halo cluster population discussed in the previous section), but that the Old Halo clusters show a significant gradient.

Correlations between globular cluster orbits and other cluster properties have been examined by van den Bergh (1993a,b). Although the majority of cluster orbits are currently poorly constrained (only a small, but growing, number of clusters have measured proper motions), van den Bergh (1993a) finds a correlation between orbital type (prograde or retrograde) and cluster Oosterhoff class. (The Oosterhoff class and dichotomy – the tendency for the periods of RR Lyrae stars to fall into one of two distinct groups – is discussed in Section 2.1 above.) This finding is interpreted by van den Bergh as indicating that one or more large gas fragments (young satellite galaxies) had the necessary age and metallicity to produce Oosterhoff I globular clusters which he finds have predominantly retrograde orbits. Further, van den Bergh (1993a) finds that clusters with horizontal branch parameter $C < 0.8$ tend to have retrograde (or radial) orbits. A comparison with Zinn's (1993) analysis shows that the majority of $C < 0.8$ clusters are identified by Zinn as being members of the Younger Halo, although Zinn's (1993) separation of the two groups is not simply made on the basis of the value of C, as explained above.

The possible significance of the group of clusters with similar metallicities and retrograde orbits described by Rodgers and Paloglou (1984) is strengthened by van den Bergh's (1993a) finding that five of the ten known clusters with retrograde orbits are constrained to the metallicity range $-1.6 < $ [Fe/H] < -1.5. He also finds that clusters at large Galactocentric distances tend to have radial orbits (see also Frenk and White 1980).

Another line of investigation pursued by several workers is the possibility that the satellite galaxies and some globular clusters of the Milky Way are associated with 'streams'. The Magellanic Stream (Wannier and Wrixon 1972; Mathewson *et al.*

1974), is the best-known example of this phenomenon. This stream consists of a long chain of gas clouds extending from the Magellanic Clouds for more than 100° and is believed to be comprised of material stripped from the Magellanic Clouds by the tidal field of the Milky Way. Two dwarf galaxies, Draco and Ursa Minor, fall in the path of this stream and are elongated along the direction of its path, supporting the idea that they are directly associated with the stream (Hunter and Tremaine 1977; Lynden-Bell 1982). There have been suggestions that similar streams pass through other dwarf galaxies surrounding the Milky Way (e.g. Majewski 1994a; Lynden-Bell 1994 and references therein).

Several attempts have been made to associate groups of halo globular clusters with similar streams. Lynden-Bell and Lynden-Bell (1995) identify several candidate streams associated with remote halo globular clusters. They point out that not all their candidate streams are likely to be genuine (partly because their technique assigns the same globulars to different candidate streams). However, the attraction of this work is that proper motion studies, which are currently being undertaken, will either confirm or rule out the reality of these streams.

Fusi Pecci *et al.* (1995) have also looked at this question and find evidence that four halo globular clusters (Pal 12, Terzan 7, Ruprecht 106 and Arp 2) all lie close to planes that pass through some of the satellite galaxies of the Milky Way. The importance of these four clusters is that they have all been shown to be significantly younger than the bulk of Milky Way halo globulars.

Perhaps the most revealing recent discovery is that of the Sagittarius dwarf galaxy which is currently being disrupted by the Milky Way (Ibata *et al.* 1994). This galaxy appears to be associated with about four globular clusters: presumably its own small globular cluster system. Even when the dwarf galaxy itself is completely torn apart by the Galaxy, these globular clusters will continue on the same orbit. Such a situation is clearly similar to the putative groups of globular clusters investigated by Lynden-Bell and Lynden-Bell (1995). Notably, one of the clusters associated with the Sagittarius dwarf, Terzan 7, is a member of the group of clusters identified by Fusi Pecci *et al.* (1995) as belonging to a satellite stream (see above).

Various inferences can be drawn from this large body of observational information and early ideas on this topic are still informative. The large spread in metallicity in the globular cluster system at a given radius (see Figure 3.6), combined with the lack of a strong metallicity gradient in the halo globular cluster system, points towards assembly of the halo from smaller fragments (Searle and Zinn 1978). This is because a considerable mass of metals must be formed in order to produce the observed metallicity spread, but the enrichment process somehow avoids the infall of metals towards the Galactic center that would lead to a strong metallicity gradient. The evidence for young stars in the halo with ages of a few Gyr (Preston *et al.* 1994 and references therein), implies a greater age spread in the field stars than the globular clusters of the halo. One possible explanation is the accretion of dwarf spheroidal galaxies by the Milky Way.

The possibility that the halo globular cluster system is comprised of two distinct populations is a significant development in this area of research. Zinn (1993) suggests that the Younger Halo globular clusters have an accretion origin, as envisaged by Searle and Zinn (1978). This is supported by the correlations found by van den Bergh

(1993a), and by the possibility that some halo clusters belong to orbital families. However, this still leaves considerable scope for the question of how these clusters actually formed. For example, they may derive from a number of gas fragments or from a single ancestral object (see van den Bergh 1993a; Zinn 1993). In the context of current galaxy formation models, the former possibility seems more likely (e.g. Silk and Wyse 1993). It is also an open question whether the ancestral objects formed stars and globular clusters before being disrupted, or whether they simply provided a source of gas to produce stars and globular clusters over an extended period of time. As we discuss in Chapter 7, one possibility is that collisions between gas clouds in the Milky Way halo may have triggered globular cluster formation. On the other hand, the discovery of the Sagittarius dwarf and its globular cluster system suggests that at least some halo globular clusters in the Milky Way originally formed around dwarf galaxies. It therefore seems likely that the halo globular cluster system, and the Younger Halo system in particular, may consist of clusters with more than one origin.

It is also possible that the Old Halo and Younger Halo globular clusters had different formation histories. Zinn (1993) argues that the most likely scenario for the Old Halo objects is formation during a relatively ordered, dissipative collapse, as outlined by several authors (e.g. Eggen *et al.* 1962; Larson 1976; Sandage 1990). Evidence in favor of such a picture includes the small age spread, spatial flattening, metallicity gradients, and kinematic properties described above. In this kind of picture, it is natural to view the disk clusters of the Milky Way as resulting from the culmination of this dissipation process. In this sense, the disk clusters and Old Halo clusters are more closely related than the clusters of the Old Halo and Younger Halo.

We note that it may be possible to produce the properties of the Old Halo clusters from conditions more similar to the Searle–Zinn picture. One can imagine, during the collapse of the protogalaxy, that gas clouds populate the halo from the Galactic center to the remote regions of the halo. The timescale for collisions and other dynamical processes in the central regions will be more rapid, so that less of an age spread will arise in globular clusters. Flattening and rotation might also be expected through cloud–cloud interactions that would tend to dissipate orbital energy of the clouds. Metallicity gradients are more problematic, as is the observation that some of Zinn's (1993) Old Halo clusters have quite large Galactocentric distances. A dissipative model for this population seems more natural, but it is probably premature to rule out the more chaotic alternative.

4

Globular cluster systems in nearby galaxies

In this chapter we discuss the properties of globular cluster systems (GCSs) around galaxies in the Local Group and slightly beyond. The proximity of such globular clusters allows more detailed information to be obtained about some of their properties than is possible for the more distant extragalactic systems discussed in Chapter 5. This information allows for comparisons between these systems and the Galactic globular cluster system. Such comparisons show that some nearby galaxies have globular clusters very much like those of the Milky Way. However, there are also galaxies which contain young, massive star clusters that have no direct counterparts in the Galaxy. Some of these objects may be young analogs of the old globular clusters in the Milky Way and other galaxies.

4.1   The M31 globular cluster system

The Andromeda Galaxy (M31) is the nearest bright spiral to the Milky Way and, consequently, its globular cluster system provides an important comparison to the Galactic globular cluster system. Many of the properties of M31 suggest that it is a slightly scaled-up version of the Milky Way, being somewhat more luminous and of earlier Hubble type. One natural question is how similar the M31 globular cluster system is to the Galactic system.

Starting with Hubble's (1932) identification of globular clusters in M31 and continuing through the present, observers have attempted to obtain a reliable globular cluster sample in this galaxy. Although globular clusters in M31 are expected to be fairly bright, surveying the population has been challenging for several reasons. If the globular clusters in M31 are like those in the Galaxy, they will be distributed over a few degrees on the sky, and have half-light radii of several arcseconds. The angular extent of the globular cluster system requires a large area survey. The typical radii of individual clusters means that they are extended in most surveys, although compact clusters can be confused with stars. Samples of globular clusters which are non-stellar suffer from contamination from background galaxies and planetary nebulae within M31. Globular clusters in M31 can only be resolved into individual stars by fairly deep, high-resolution imaging, which is not typical of surveys. Redshift surveys address most of these issues directly, but are time consuming. Perhaps the most serious problem for obtaining a complete sample is the bright and variable surface brightness of M31 against which many clusters are projected.

Most of these difficulties have been overcome through a number of observational

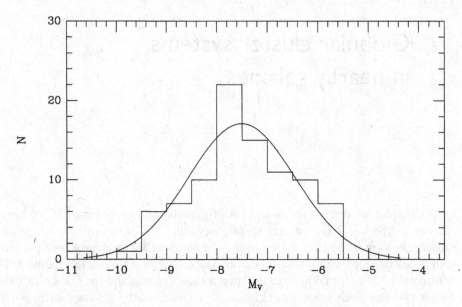

Fig. 4.1. The V-band luminosity distribution of globular clusters in M31. Also shown is the best-fitting Gaussian to the data.

programs (reviewed by Fusi Pecci *et al.* 1993a), and much is now known about the M31 globular cluster system. In order to compare the GCSs of M31 and the Milky Way, we follow the general scheme of Chapter 3 in describing the various facets of the M31 globular clusters. However, since more is known about the M31 *system* than its individual globular clusters, we begin by comparing the systemic properties of the globular clusters in the Milky Way and M31.

4.1.1 The luminosity distribution

The total number of globular clusters in M31 is around 450 ± 100 objects, or a factor of 2.5 greater than the Milky Way system (e.g. Battistini *et al.* 1993). Reliable photometry is only available for a subset of the known candidates, but the 81 clusters in the sample compiled by Reed *et al.* (1994) are more than adequate to construct a useful luminosity function. As discussed in Chapter 3, there are various ways of presenting such a luminosity function. Figure 4.1 shows the traditional histogram in terms of globular cluster magnitude where the best-fitting Gaussian has been superimposed. Absolute magnitudes are taken from Reed *et al.* (1994) and are based on a true distance modulus to M31 of 24.3 and a reddening of $E(B - V) = 0.11$. Uncertainties in the M31 distance modulus and reddening corrections lead to an uncertainty in the derived absolute magnitudes of $0.1 - 0.2$ mag. Note that the lack of clusters fainter than $M_V = -5.5$ is a consequence of incompleteness.

The best-fitting Gaussian to the M31 globular cluster luminosity distribution has $M_V = -7.51$ $(-7.35, -7.69)$ and $\sigma = 1.05$ $(0.94, 1.21)$ (Ashman *et al.* 1994; see also Secker 1992). Values in parentheses represent bootstrapped 90% confidence intervals. Thus the peak of the M31 luminosity function is slightly brighter than that

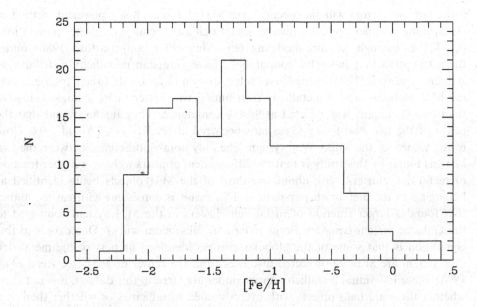

Fig. 4.2. The metallicity distribution of globular clusters around M31.

of the Milky Way, and the dispersion is narrower. However, these differences are not statistically significant (Ashman *et al.* 1994), even ignoring additional distance and reddening uncertainties. Indeed, the similarity between the globular cluster luminosity function from galaxy to galaxy, explored in detail in Chapter 5, represents an important constraint on the formation and evolution of globular cluster systems.

If the M31 luminosity function is plotted in terms of luminosity rather than magnitude, the form is, unsurprisingly, also similar to that of the Milky Way. The slope of the best-fitting power law to the bright end of the distribution has a value around -1.8, and the power-law break occurs around 10^5 $L_\odot$ (McLaughlin 1994).

4.1.2   *Metallicity and kinematics: a disk–halo division?*

The metallicity distribution of M31 globular clusters is shown in Figure 4.2 based on the spectroscopic metallicities of Huchra *et al.* (1991). This sample has 144 clusters and excludes six objects of the Huchra *et al.* (1991) list that were flagged as possible members of the globular cluster system of NGC 205. The metallicity histogram exhibits an asymmetric form, with a tail of high-metallicity clusters. A comparison of Figure 4.2 with the similar histogram presented earlier for the Milky Way system (Figure 3.4) illustrates that the distinction between metal-rich and metal-poor clusters in the Galaxy is not immediately apparent in the M31 system. However, the larger errors associated with the metallicity determinations of the M31 clusters could smear out evidence for two distinct populations, so a more quantitative analysis is called for.

In the Milky Way, the metallicity distributions of both the halo and disk clusters are individually well fit by Gaussian distributions. This may be partly due to random errors tending to produce a Gaussian distribution rather than an intrinsic Gaussian form, but such errors will also occur in the M31 dataset. Thus a powerful method of establishing whether there are distinct metal-rich and metal-poor cluster populations in M31 is through mixture-modeling (e.g. McLachlan and Basford 1988) under the assumption that individual populations have Gaussian metallicity distributions. Ashman and Bird (1993) carried out such a study and found that the hypothesis that the M31 globular cluster metallicity distribution was better fit by a single Gaussian than two Gaussians was rejected at 98.4% confidence. They further found that the peaks of the two best-fitting Gaussians occurred at [Fe/H] = −1.5 and −0.6, close to the values for the Milky Way system. The only notable difference between the two systems found by this study is that the M31 system appears to have a higher fraction of metal-rich clusters, with about one-third of the M31 objects being identified as belonging to the metal-rich population. This result is consistent with earlier studies that found a larger fraction of metal-rich clusters in the M31 system compared to the Galactic system (van den Bergh 1969 and subsequent work). One caveat to this conclusion is that some of the globular clusters identified as part of the metal-rich disk system are at large projected distances from the center of M31 (see Reed *et al.* 1994). Since individual metallicity uncertainties are large in this dataset, it is not clear whether these are halo objects with overestimated metallicities or whether there is a disk system which is more extended than that in the Milky Way.

While the evidence for two populations of M31 globular clusters is suggestive of a disk–halo division similar to that in the Milky Way, kinematic evidence is also required to demonstrate such a division definitively. The difficulty is that most of the current velocities for M31 globular clusters have relatively large errors so that, along with projection effects, it is challenging to clearly identify a rotating disk system. Despite these limitations, Huchra *et al.* (1991) did find some evidence that the metal-rich clusters in M31 constitute a rotating system. The inner metal-rich clusters are rotating at around 100–200 km/s, with the value dropping to around 60 km/s if all metal-rich clusters ([Fe/H] > −0.8) are included (Huchra *et al.* 1991; see also Huchra *et al.* 1982). Based on the metallicity division obtained from mixture-modeling, Ashman and Bird (1993) derived a halo rotation velocity of 39 km/s, although the value is statistically consistent with zero. Huchra *et al.* (1991) use the globular cluster velocities to probe the mass distribution of M31 and find strong evidence for a dark matter halo (see also Federici *et al.* 1990, 1993).

There is some evidence for a weak metallicity gradient in the M31 globular cluster system, but Huchra (1993) suggests this is produced by the more rapidly rotating metal-rich clusters. In other words, if there is a disk–halo division in the M31 system then, like the Milky Way, the halo clusters of M31 show no evidence of a metallicity gradient. Another similarity between the M31 and Milky Way systems is that the M31 globular clusters show no correlation between metallicity and luminosity (Huchra *et al.* 1991; Covino and Pasinetti Fracassini 1993).

4.1.3   Other properties

The radial distribution of the M31 globular cluster system is well-fit, at least outside the central regions, by a single power law. The volume density profile is $\rho \propto R^{-3}$ (Racine 1991; Fusi Pecci *et al.* 1993a), corresponding to a surface density profile scaling like r^{-2}, where r is the projected distance from the center of M31. The form of this power law is consistent with the best fit to the radial distribution of Milky Way globulars. However, the stellar light of the M31 halo follows a much steeper power law of $\rho \propto R^{-5}$ in the outer regions, with an exponent around −4 closer to the center of the galaxy (Pritchet and van den Bergh 1994). In other words, the M31 globular cluster system appears to be more spatially extended than the halo light of M31, whereas, in the Milky Way, the globular clusters and halo stars follow the same distribution. One potential concern is that it may be more difficult in M31 to define a pure sample of halo stars from which to derive the stellar profile. Another worry is possible incompleteness in the globular cluster sample at large projected distances from the center of M31.

The M31 globular cluster distribution can also be fit by other functions including a de Vaucouleurs law (Battistini *et al.* 1993). The effective radius of the best-fitting de Vaucouleurs profile is greater for the globular cluster system than the halo stars, confirming the more extended nature of the globular cluster system, with the same caveat concerning bulge contamination mentioned above. Battistini *et al.* (1993) also find that there is a deficit of clusters in the inner regions of M31 relative to the best-fitting de Vaucouleurs profile. Part of this deficit may be a result of the difficulty in detecting concentrated clusters in the central regions of M31. As we discuss in Chapter 5, a central deficit of clusters is a common feature of globular cluster systems.

Like the Milky Way and a handful of other galaxies where such data are available, the radii of globular clusters in M31 increase with projected distance from the galaxy center (Crampton *et al.* 1985). As discussed in the previous chapter and in Section 4.2 below, a plausible explanation for this trend is that denser environments produce more concentrated globular clusters (van den Bergh 1991a).

Finally, there is some evidence that there may be 'groups' of globular clusters in M31 with similiar orbital and/or spatial characteristics, much like those tentatively identified in the Milky Way (see Section 3.6). Ashman and Bird (1993) found evidence for correlations between positions and velocities for globular clusters in their halo subsample. The nature of this globular cluster clustering is unclear, but it could be produced either by dynamically associated groups of globular clusters or by families of globular clusters following similar orbits. In either case, the reason for the physical association of these clusters may be that they were originally the globular cluster systems of dwarf galaxies that have been accreted by M31.

4.1.4 *Observations of individual globular clusters*

The stellar content of M31 clusters has been investigated spectroscopically and, more recently, through color–magnitude diagrams. Spectroscopic studies have established one of the most marked differences between M31 and Milky Way globular clusters: CN abundances in M31 clusters are much higher than in their Milky Way counterparts (Burstein *et al.* 1984; Tripicco 1989; Davidge 1990; Brodie and Huchra 1990, 1991). The effect appears to be the most notable for metal-rich clusters. The origin of this difference is unclear, although Tripicco (1993) suggests that it may be related to the CN bimodality in Milky Way globular clusters (see Section 2.3). If this is the case, the globular clusters in M31 have a higher proportion of CN-rich stars than their counterparts in the Milky Way.

Other spectral differences between globular clusters in the two galaxies are more controversial. There have been several suggestions that Balmer line strengths in M31 globulars are stronger than those of Milky Way clusters at the same metallicity (Spinrad and Schweizer 1972; Burstein *et al.* 1984). Follow up studies suggested that this was probably not a general property of M31 globulars, but that some clusters did have enhanced Balmer line strengths (Brodie and Huchra 1990, 1991; Huchra *et al.* 1991). The line strengths are indicative of a hot stellar population which has sometimes been interpreted as evidence for some M31 clusters being significantly younger than Milky Way globulars. Early UV studies supported the presence of bluer horizontal branches in M31 clusters (Cacciari *et al.* 1982; Cowley and Burstein 1988), but more recent work indicates that UV spectra alone do not provide evidence for a stellar population difference between globular clusters in the Milky Way and M31 (Crotts *et al.* 1990). A review of these issues is provided by Tripicco (1993).

Comparisons of other properties of M31 and Milky Way globular clusters primarily point to similarities. Djorgovski *et al.* (1997) have constructed fundamental plane relations for M31 globulars analogous to those described in Section 3.2 for the Milky Way system. Trends between mass-to-light ratio, velocity dispersion, luminosity, and other key variables are indistinguishable from the Milky Way relations.

HST observations are now resolving stars fainter than the horizontal branch in M31 globular clusters (e.g. Rich *et al.* 1996), thereby allowing the colors of the giant and horizontal branches to be studied in these objects and compared with Milky Way globular clusters. Early results suggest that the M31 globular clusters are much the same as Milky Way globular clusters in this respect. An important next step in these studies is to look for correlations between horizontal branch morphologies and metallicities to investigate second parameter effects (see Section 2.1). As discussed in Section 3.1, recent observations by Ajhar *et al.* (1996) and Fusi Pecci *et al.* (1996) have provided an important calibration of the [Fe/H]–M_V(RR) relation for M31 clusters that constrains current uncertainties in globular cluster distances within the Milky Way.

A collapsed-core globular cluster has been detected in M31 (Bendinelli *et al.* 1993; Grillmair *et al.* 1996), and it is anticipated that current and future *HST* observations will reveal more. A comparison of the spatial distribution of such clusters in M31 and the Milky Way would be extremely valuable for understanding the importance of external dynamical effects on producing collapsed-core clusters (see Section 3.5). Grillmair *et al.* (1996) have derived surface brightness profiles for a small sample of M31 clusters from *HST* imaging. Other than the collapsed-core cluster mentioned above, the remaining objects are well fit by King models (see Section 2.6). In the outer regions of these clusters, Grillmair *et al.* (1996) find tentative evidence for tidal tails which are also observed in some Milky Way globulars. Fusi Pecci *et al.* (1994) have presented results for 13 M31 globular clusters based on pre-refurbishment *HST* images, and conclude that the structural parameters of globular clusters in their sample are indistinguishable from those of Milky Way globulars.

4.2   The Magellanic Clouds

The Large and Small Magellanic Clouds (LMC and SMC respectively) are the nearest galaxies to the Milky Way. Both are gas-rich, dwarf irregular galaxies that are satellites of the Galaxy. They are closer to the Galactic center than a few

Milky Way globular clusters, with Galactocentric distances of roughly 50 kpc. The Magellanic Clouds are therefore located within the dark halo of the Milky Way according to current estimates of the extent of this halo (Ashman 1992).

4.2.1   Old globular clusters

Traditionally, globular clusters in the Magellanic Clouds have been defined as massive star clusters that are manifestly old (ages greater than around 10 Gyr). Even with this definition, there has been some debate over which old star clusters in the Magellanic Clouds are genuine globulars. The two principal difficulties are establishing whether a cluster is associated with one of the Magellanic Clouds or with the Galactic population of globular clusters, and whether an individual cluster is a massive, old open cluster or a genuine globular cluster. The second issue seems to be largely a problem of semantics. The low masses and evidence for an age spread among some halo globular clusters of the Milky Way suggest that there is some overlap in the properties of old open and globular clusters. The question of membership can be settled to some extent by radial velocity measurements of cluster stars.

With these caveats, there seems to be a general consensus that the SMC has just one old (> 10 Gyr) globular cluster, NGC 121 (van den Bergh 1991a; Harris 1991, and references therein). The estimated number of globular clusters in the LMC is currently 13 (Suntzeff *et al.* 1992), with the possibility that there are additional clusters yet to be discovered. Clearly, nothing can be said about the SMC globular cluster system on the basis of one object, so for the remainder of this subsection we focus our attention on the LMC.

The similarities and differences between the LMC globular cluster system and the systems of other galaxies have been reviewed by van den Bergh (1991a, 1993b). The luminosity function of the old LMC globulars is consistent with that of the Milky Way (see also Suntzeff *et al.* 1992), although the small number of objects limits the usefulness of this comparison. The mean magnitude of LMC globulars is indistinguishable from that of the Milky Way system. The mean metallicity of the 13 clusters considered by Suntzeff *et al.* (1992) is [Fe/H] $\simeq$ −1.8, lower than the mean metallicity of the Milky Way globular cluster system, but similar to the mean metallicity around −1.6 of the Milky Way *halo* clusters (Suntzeff *et al.* 1992; Ashman and Bird 1993). No metallicity gradient is observed in the LMC globular cluster system (Da Costa 1993 and references therein).

The color–magnitude diagrams of old LMC globulars also appear to be similar to their Milky Way counterparts. Recent *HST* observations by Mighell *et al.* (1996) of the LMC globular cluster Hodge 11 yield an age that is indistinguishable from the Milky Way globular M92, with an uncertainty in the relative ages of no more than 21%. *HST* observations of other LMC clusters are expected to yield more relative ages with smaller uncertainties. Earlier results on the stellar content of the old LMC globulars have been summarized by Da Costa (1993). Important findings include the high degree of chemical homogeneity in two LMC globular clusters (Suntzeff *et al.* 1992) and the possibility that the LMC globulars exhibit the same second parameter effect observed for Milky Way halo globular clusters (see Sections 2.1 and 3.6). Using color–magnitude diagrams from Walker (1990, 1992a,b) and abundances from Olszewski *et al.* (1991), Da Costa (1993) concludes that the LMC globular

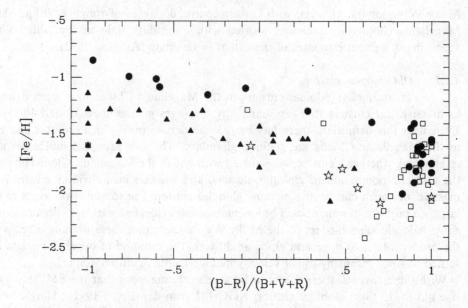

Fig. 4.3. A plot of metallicity against horizontal branch morphology for old LMC globular clusters (represented by star symbols). Also shown are the halo globular clusters of the Milky Way, where the symbols have the same meaning as in Figure 3.11. (From Da Costa (1993) and data supplied by R. Zinn.)

clusters are similar to the Milky Way clusters of Zinn's (1993) Younger Halo (see Section 3.6). Specifically, Da Costa (1993) plots the horizontal branch parameter C against [Fe/H], as shown in Figure 4.3. The LMC clusters fall off the mean C–[Fe/H] relation of the Old Halo clusters, as do the Younger Halo globular clusters of the Milky Way. The old globular cluster in the SMC, NGC 121, shares this property (Da Costa 1993).

The half-light radii of LMC clusters are significantly larger on average than the Milky Way globulars (Bhatia and MacGillivray 1988; van den Bergh 1991a). This conclusion is well established for young LMC and SMC clusters which have $r_{1/2}$ values typically around three times larger than Milky Way globulars. The difference also seems to hold when old globular clusters in the LMC are compared to Milky Way globulars, although the information on $r_{1/2}$ for LMC globulars is limited. Suntzeff *et al.* (1992) also discuss the related issue of the tidal radii of LMC globular clusters, and find that r_t increases with distance from the center of the LMC, as is the case for the Milky Way, M31 and NGC 5128. The difference in size between the old LMC globular clusters and those in the Milky Way, as well as the trend between radius and galactocentric distance, may simply indicate that globular clusters that form in lower-density environments tend to be more diffuse (see van den Bergh 1991a). The LMC is of lower density than the inner halo of the Milky Way where the majority of Galactic globular clusters are found. We discuss the physical origins of the relation between environmental density and globular cluster density in Chapter 7.

There is limited information on the internal dynamics of some of the old LMC globular clusters (Meylan *et al.* 1989; Suntzeff *et al.* 1992). These results indicate that

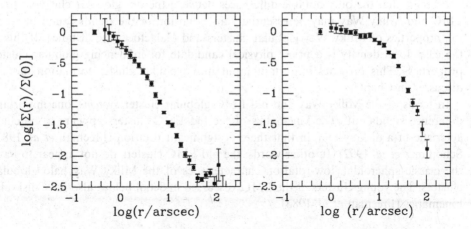

Fig. 4.4. *B*-band surface brightness profiles of the LMC clusters NGC 2019 (left) and NGC 1978 (right). The surface brightness has been normalized to the value at the innermost point for each cluster. (From data supplied by G. Meylan.)

dynamical masses and mass-to-light ratios for these clusters are comparable to their counterparts in the Milky Way (see also Dubath *et al.* 1990; Meylan *et al.* 1991; Dubath *et al.* 1993; Mateo 1993). The surface brightness profiles of LMC globulars are similar to those of Milky Way clusters, exhibiting a range of concentrations when fit by King models, as well as core-collapsed clusters (e.g. Meylan and Djorgovski 1987). In Figure 4.4 we show surface brightness profiles of two LMC clusters studied by Meylan and Djorgovski (1987). NGC 2019 has the classic profile of a core-collapsed cluster, whereas NGC 1978 has a flat central profile. (Note that NGC 1978 is not included in the 13 old globular clusters considered by Suntzeff *et al.* (1992), but is classified by Meylan and Djorgovski (1987) as an old cluster.)

Perhaps the most intriguing difference between Milky Way and LMC globular clusters is that the LMC objects are considerably more flattened (e.g. van den Bergh and Morbey 1984). This difference persists for the young star clusters in the two galaxies in addition to the old globular clusters. The origin of this difference is not understood. It is believed that the less dramatic flattening of a few Milky Way globulars is probably due to rotation (e.g. Pryor *et al.* 1986), which would suggest that the LMC clusters are rotating more rapidly, but this primarily shifts the problem from shape to angular momentum. It has been noted by van den Bergh (1991a) that the differences in size and flattening between globular clusters in the LMC and Milky Way might be related phenomena. He suggests that the physical connection may be that low-density clusters are more easily distorted and thus more likely to be flattened. An analysis of the Milky Way globular clusters does not show any clear trend between $r_{1/2}$ and flattening, as might be expected if this intepretation were correct, although van den Bergh notes that two of the four highly flattened Milky Way globulars have large $r_{1/2}$. It has also been pointed out by van den Bergh (1996) that the most luminous globular clusters in the Milky Way, M31 and the LMC are considerably more flattened than typical clusters in these galaxies.

It seems that the only notable differences between the old globular clusters of the LMC and Milky Way may be essentially an extension of the trends found between the properties of Milky Way globular clusters and Galactocentric distance. If this is the case, local density is a prime physical candidate for influencing globular cluster properties. This provides important input into globular cluster formation models discussed in Chapter 7.

In terms of the Milky Way and old LMC globular cluster *systems*, one important difference stands out. The kinematics of the old LMC clusters appears to be characteristic of a disk system, in that there is significant rotation (Freeman *et al.* 1983; Schommer *et al.* 1992). In other words, the old LMC clusters do not appear to have the classic spheroidal, low-rotation characteristics of the Milky Way halo globular clusters. The young clusters of the LMC, discussed in detail below, also have disk-like kinematics (Freeman *et al.* 1983).

4.2.2   *Young globular clusters?*

Both the LMC and SMC have large populations of young ($\lesssim 1$ Gyr) and intermediate-age (1 Gyr to $\lesssim 10$ Gyr) star clusters. The properties of these clusters are discussed at length in papers in the conference proceedings edited by Haynes and Milne (1991), and by several other authors including van den Bergh (1993b), Da Costa (1993), and Richtler (1993), where references to earlier work can also be found.

There has been considerable discussion over the years that some of these star clusters might be younger analogs of the old globulars found around other galaxies. The question has recently received renewed interest and importance with claims that young globular clusters are currently being formed in galaxy mergers and other environments (see Chapter 5). Part of the resistance to the idea of young globular clusters in the Magellanic Clouds and elsewhere may spring from the traditional view that globular clusters are, by definition, old objects. We feel that if a young star cluster will evolve, over the next 10 Gyr or so, to resemble old globular clusters in the Milky Way and other galaxies, then such an object can be meaningfully described as a young globular cluster. There are two problems with this operational definition. One is that it requires a detailed knowledge of various cluster properties to predict how a cluster will evolve. With only an age estimate from colors, there are large uncertainties arising from the unknown stellar mass function. Even if the cluster mass is determined dynamically, there are still uncertainties about the cluster's orbit in its host galaxy and its ability to survive destructive dynamical effects. A second problem is that the distinction between old globular clusters and old open clusters is not completely clear-cut.

It is quite evident in the LMC and SMC that the vast majority of young clusters are *not* globular clusters. Not only is the inferred mean mass significantly lower than that of Milky Way globular clusters, but the bulk of the young clusters in the Magellanic Clouds are likely to dissolve over the next few Gyr (e.g. Hodge 1987; van den Bergh 1991a). Further, the luminosity function of young clusters in the LMC can be well fit by a power law (e.g. van den Bergh and Lafontaine 1984; Elson and Fall 1985), whereas the luminosity function of old globular clusters in the Milky Way and other galaxies has the log-normal form described in Chapters 3 and 5. This difference has been used to argue against the possibility that *any* of the young clusters in the LMC (and, by extension, the SMC) are young *globular* clusters. However, Mateo (1993) has pointed out that the luminosity function of young LMC clusters can be fit by a superposition of a power-law and a log-normal distribution, the idea being that the young clusters are comprised of both open and globular clusters (see also Ashman and Zepf 1992). Indeed, there is a 'bump' in the LMC cluster luminosity function that Mateo (1993) tentatively identifies as being due to young globular clusters. Open clusters outnumber globular clusters by around twenty to one in this model. It should be stressed that this model, and the young cluster luminosity function, can *not* be regarded as requiring the presence of young globular clusters, but it is apparent that one cannot rule out the presence of young globular clusters in the LMC and SMC simply on the basis of the power-law luminosity function.

Perhaps the central question that stems from the power-law luminosity function of young clusters is the form of the evolved luminosity function. Specifically, after 10 Gyr or so, will the cluster luminosity function resemble that of old globulars? This is necessary if we are to identify some of the young clusters with globular clusters. The question is not an easy one to answer, since a solution requires a detailed understanding of the evolution and destruction of the star clusters. A more effective strategy may be provided by other recently discovered young cluster systems containing globular cluster candidates (see Chapter 5). However, it is generally accepted (see Section 3.5) that low-mass clusters are more likely to dissolve, which qualitatively acts in the right sense to evolve a power-law luminosity function into a log-normal function. Elson *et al.* (1987) found evidence that many LMC clusters exceed their tidal radii, implying that the clusters are losing mass and suggesting that such clusters may be in the process of dissolving.

4.2.3 The formation and evolution of the Magellanic Clouds

Like other cluster systems, the young and old star clusters of the Magellanic Clouds provide insights into the formation and evolution of their host galaxies. In terms of the old globular clusters in the LMC, the disk-like kinematics are intriguing, perhaps suggesting that some star formation occurred *before* these clusters formed. (This is in contrast to the situation in the Milky Way where the evidence suggests that the halo globulars formed during the initial epoch of star formation.) The lack of a metallicity gradient in the old LMC globular cluster system argues against slow, dissipational collapse in which the metallicity is increased through enrichment, but with the small number of clusters this conclusion is not firm. Further, the disk-like nature of the system argues *for* dissipation.

The intermediate-age and young clusters have provided more information on the evolution of the Magellanic Clouds. The most striking aspect of the LMC system is an almost complete lack of clusters in the age range between about 3 and 12 Gyr (e.g. Mateo *et al.* 1986; Da Costa 1991; Olszewski 1993). The SMC does not exhibit such an age gap. While the possibility of globular cluster destruction complicates the intepretation of these results, the current consensus seems to be that the LMC formed star clusters (including globular clusters) around 12 Gyr ago, experienced a quiescent period, then began a second phase of cluster formation 3 Gyr ago which continues to the present day. The SMC, on the other hand, seems to have formed star clusters at a continuous, but more leisurely rate, over its entire history. Models which attempt to explain the cluster formation history through interactions between the Clouds stumble over the observation that the LMC shows two distinct cluster formation bursts, whereas the SMC does not. Nevertheless, it seems reasonable that some kind of disturbance is required to 'turn on' cluster formation in the LMC.

Finally, the kinematics of the LMC clusters have been used to analyze the mass distribution of this galaxy. By combining cluster velocities with velocities of planetary nebulae and HI measurements, Schommer *et al.* (1992) concluded that it was difficult to explain the LMC rotation curve purely by visible material. Thus, like other galaxies, it seems that the LMC is surrounded by a dark halo. As we discuss in Chapter 5, globular clusters around more distant galaxies provide valuable dynamical tracers that can be used to probe the dark halos of such galaxies.

4.3   Dwarf spheroidals

The most numerous galaxies in the Local Group are the faint, low surface brightness dwarf spheroidals. Nine such galaxies are generally regarded as satellites of the Milky Way, but only two have detected globular clusters systems. It has been known for decades that the Fornax dwarf spheroidal has a handful of globular clusters (Baade and Hubble 1939; Hodge 1961, 1965). In contrast, the Sagittarius dwarf was only recently discovered (Ibata *et al.* 1994), and seems to be associated with four globular clusters previously regarded as part of the Milky Way system.

With such small numbers, it is clearly difficult to draw many conclusions about the globular cluster systems of these two galaxies. Nevertheless, several interesting results have been obtained. The Fornax system consists of five or six clusters (the sixth is rather faint and not always included in the system) which have a stellar content much like that of halo globulars in the Milky Way (e.g. Buonanno *et al.* 1985). Several studies of the metallicities of these clusters have been carried out (e.g. Zinn and Persson 1981; Brodie and Huchra 1991; Dubath *et al.* 1992) revealing that the system is metal-poor, but that one cluster appears to be nearly an order of magnitude more metal-rich than the others (see also Beauchamp *et al.* 1995). The combination of small numbers and an asymmetric metallicity distribution means that appropriate statistics must be used in deriving a mean metallicity. Ashman and Bird (1993) used a bootstrap resampling technique and derived a mean metallicity of $[Fe/H] = -1.91$ $(-1.28, -2.09)$, where values in parentheses represent the 90% confidence interval on the mean. This interval brackets the mean metallicity of the Milky Way halo globulars and that of the old LMC globulars.

More recent work on the color–magnitude diagrams of Fornax globular clusters has revealed that these objects are similar to the 'Younger Halo' clusters of the Milky Way identified by Zinn (1993; also Section 3.6) and the old globulars of the LMC (Beauchamp *et al.* 1995; Smith *et al.* 1996). The two clusters studied by Smith *et al.* (1996) exhibit a second parameter effect (see Section 2.1), having similar metallicities but different horizontal branch morphologies. Both these clusters have redder horizontal branches than any Galactic globulars at the same metallicity. It is these very red horizontal branches that suggest a connection to the Younger Halo cluster population and the red horizontal branches of LMC globular clusters described above.

A study of the structural properties of the Fornax globular clusters has been carried out by Rodgers and Roberts (1994). Two of the clusters have surface brightness profiles well fit by King models, as is the case for the majority of Galactic globulars. The remaining three clusters, however, are more extended in their outer regions and cannot be fit by a King model. Rodgers and Roberts (1994) speculate that the lack of a strong tidal field to truncate the Fornax clusters may be responsible for this structural characteristic.

The recent discovery of the Sagittarius dwarf means that less is known about its globular cluster system, but Ibata *et al.* (1995) present a strong case that the clusters M54, Arp 2, Terzan 7 and Terzan 8 are associated with this galaxy rather than the Milky Way. Metallicity estimates are available for these clusters (Da Costa and Armandroff 1995) and reveal the curious fact that, like Fornax, one of the clusters, Terzan 7, is about an order of magnitude more metal-rich than the rest. This suggests that even in the smallest globular cluster systems an appreciable metallicity spread between globular clusters is possible (and perhaps usual). This finding has some interesting implications for models of globular cluster formation (Chapter 7).

The other dwarf galaxies in the Local Group with known globular cluster systems are the dwarf *ellipticals* NGC 147, NGC 185 and NGC 205 (see Harris *et al.* 1991 and references therein) that are believed to be satellites of M31. These galaxies are distinct from the dwarf spheroidals discussed above, being more compact and having higher surface brightnesses. The GCSs contain from two to a few known members. The globular clusters are generally metal-poor and show no obvious differences from the globular clusters in the Fornax and Sagittarius dwarfs.

Given that the majority of the Local Group dwarf galaxies do *not* have globular cluster systems, one might ask what is special about those that do. It seems that the most likely explanation is that Fornax, Sagittarius, and the M31 satellites are more luminous than the dwarfs that do not have globular cluster systems. These galaxies have only managed to produce a few clusters each, so it is not surprising that even fainter galaxies did not produce any globular clusters at all. In fact, as we discuss in the next chapter, Fornax and Sagittarius seem to have been relative efficient at forming globular clusters compared to other types of galaxy. Specifically, Fornax has produced as many globular clusters per unit stellar luminosity as the most populous GCSs around galaxies such as M87 (see Chapter 5).

4.4   M33 and M81

M33 is a late-type, low-luminosity spiral galaxy, sometimes regarded as a transitional object between large spirals such as the Milky Way and M31 and irregular galaxies like the Magellanic Clouds. In this context, its well-studied cluster system provides important links between the other globular cluster systems of the Local Group.

A list of photometric studies and surveys of clusters in M33 is provided by Christian (1993). One important issue in discussing the cluster system is that, like the Magellanic Clouds, M33 contains a number of young and intermediate-aged, concentrated star clusters. We do not address the question of whether some of these younger clusters are analogs of classical old globular clusters since the arguments for and against the idea are similar to those for the Magellanic Cloud clusters discussed above. There is far more detailed information about young clusters in the Magellanic Clouds compared to the situation in M33 due to the proximity of the former objects. If some of the young clusters in the Magellanic Clouds are genuine young *globular* clusters, by extenstion it is natural to suppose that some of the young M33 clusters also fall into this category.

Less controversial is the presence of a system of typical old globular clusters around M33. Photometric studies have identified around 25 such clusters (Christian and Schommer 1988 and references therein) which probably constitute most of the old globular clusters in this galaxy. These studies also indicate color distributions that have a similar spread to the Milky Way system, although the small number of objects makes a detailed comparison impossible. The mean metallicity of the globular clusters is found to be [Fe/H] $\simeq -1.6$, with some evidence for a metallicity gradient (Schommer 1993, and references therein). Planned *HST* observations to obtain color–magnitude diagrams of M33 globular clusters will greatly add to our understanding of the stellar content of these objects.

The construction of a reliable luminosity function is also limited by small numbers, as well as by the possibility that some of the old clusters are misclassified objects of intermediate age. However, there is some evidence that the peak of the luminosity function might be somewhat fainter than in the Milky Way and M31 (Christian and Schommer 1988). It has been noted that, unlike the Milky Way and M31, the old cluster system of M33 does not extend much beyond the disk of this galaxy. This may provide evidence that M33 has a less extended halo than the larger spirals, even when appropriate scaling for its smaller disk is made. However, it is worth pointing out that if the total number of globular clusters in M33 is not much greater than the known population, a surface density profile of the same form as other galaxies implies that the expected number of globulars beyond the disk is small.

There is some kinematic information on the M33 globular cluster system (Cohen *et al.* 1984; Schommer *et al.* 1991; Schommer 1993). The primary conclusion is that the old M33 globulars represent a typical halo distribution with little net rotation. In this regard they are similar to the globular clusters in M31 and the Milky Way. As mentioned above, the old globular clusters of the LMC do *not* share this property, exhibiting instead disk kinematics. In this regard, M33 appears to be more closely related to the large spirals.

Although M81 is beyond the Local Group, it is sufficiently nearby (around 3 Mpc) that observations of its globular cluster system require a rather different strategy than the more distant systems discussed in the next chapter. Specifically, this large spiral galaxy is too distant for its globular clusters to be easily resolved with ground-based imaging, but it is close enough that it covers a large area on the sky, so that detecting globular clusters amongst the many candidate sources is a difficult task.

Despite these problems, Perelmuter and Racine (1995) have carried out a photometric survey for candidate globular clusters in M81 and have obtained around 70 good candidates. They estimate that the total population of globular clusters is around 210. Perelmuter and Racine (1995) find that the projected surface density profile of the system follows the r^{-2} form found for the two large Local Group spirals, and that the color (and, by assumption, metallicity) distributions of the three globular cluster systems are similar. The globular cluster luminosity function of the M81 GCS also appears to be similar to that found for the Galaxy and M31 (Perelmuter and Racine 1995). Using the data of Perelmuter and Racine (1995) and a distance modulus of 27.8, Blakeslee and Tonry (1996) find a peak magnitude around $M_V \simeq -7.8 \pm 0.4$. This is marginally consistent with the peak values found for the Galaxy and M31 within the large uncertainties.

The metallicity and kinematics of the M81 system have been analyzed by Brodie and Huchra (1991) and Perelmuter *et al.* (1995). The latter study presents spectra for 25 globular clusters and yields a mean metallicity estimate for the globular clusters of $[Fe/H] = -1.48 \pm 0.19$. Since the clusters in this sample are generally not observed close to the center of M81, this value is probably most usefully compared with the metallicity of the *halo* clusters in M31 and the Milky Way. All three values are remarkably similar. The velocities of the globular clusters in M81 provide evidence that this galaxy has a massive dark halo (Perelmuter *et al.* 1995).

4.5 Other irregular galaxies

There are a number of irregular galaxies at distances slightly beyond the Local Group whose globular cluster systems have also been studied in detail. Several of these have attracted particular attention because they are known to have 'super star clusters', which are very luminous, compact clusters of young stars. The luminosities of these super star clusters exceed that of the brightest star cluster in the Local Group (R136a in the LMC) by several magnitudes. Because of their high luminosities, the existence of super star clusters in several nearby irregular galaxies has been known for some time (e.g. Kennicutt and Chu 1988; Arp and Sandage 1985, and references therein). The nearby galaxies in which super star clusters have been found include NGC 1569, NGC 1705, NGC 253, NGC 5253, and M82. All these galaxies are undergoing starbursts, and most of them are classified as amorphous or irregular.

Ground-based imaging established the bright luminosities and blue colors of these super star clusters, but was unable to resolve the objects. At the distance of these galaxies, ground-based images only limited the objects to sizes of less than several tens of parsecs, so it was not clear that these objects were as compact as young LMC clusters or Galactic globular clusters. *HST* imaging has answered this question definitively, showing that the super star clusters have sizes of a few parsecs, just like their possible globular cluster counterparts in the Local Group (e.g. O'Connell *et al.* 1994 for NGC 1569 and NGC 1705; O'Connell *et al.* 1995 for M82; Watson *et al.* 1996 for NGC 253).

Another challenge is to determine the age of these clusters. Combined with the observed luminosity, the age is particuarly useful because it allows an estimate of the mass of the system through models of stellar populations which give a mass-to-light ratio at a given age. This area has been most thoroughly pursued for the super star clusters NGC 1569-A and B and NGC 1705-A, for which there are both colors and moderate resolution spectroscopy (Arp and Sandage 1985 for NGC1569-A and B; Melnick *et al.* 1985 for NGC 1705-A). Based on broad-band colors and these spectra, O'Connell *et al.* (1994) estimate an age of ~ 15 Myr for these clusters. With this age and the observed luminosities, the estimated masses for these clusters are in the range from 10^5 to 10^6 $M_\odot$, very similar to those of the most massive young LMC clusters, and to old Galactic globular clusters.

The similarity of the properties of these objects to those expected of Galactic globular clusters at very young ages is striking. One of the uncertainties in the comparison is the mass estimate, which depends on an adopted stellar initial mass function which is not observed. A potentially more direct way to measure the cluster mass is through dynamical means. Although individual stars have not been resolved in massive clusters in galaxies more distant than the LMC, integrated velocity dispersions are feasible in some cases. Unfortunately, the integrated velocity dispersions of very young clusters ($\sim$ 10 Myr) do not provide unambiguous masses. The problem is that the signal in the lines from which the dispersion is derived is dominated by a small number of massive stars. Thus there are potential problems in obtaining a completely fair velocity sample from such observations, particularly since clusters of such a young age may not be fully relaxed. Despite this caveat, the discovery that the integrated velocity dispersions of NGC 1705-A and NGC 1569-A and B give masses in the range given above (Ho and Filippenko 1996a,b) lends weight to the arguments that these objects are young analogs of Galactic globular clusters.

The observations of objects which may be young globular clusters is a direct challenge to the traditional view that all globular clusters are old. In most nearby galaxies displaying this phenomenon, only a handful of potential young globular clusters are observed. In the next chapter, we discuss systems of candidate young globular clusters in slightly more distant galaxies. In some of these systems, hundreds of candidate young globular clusters have been detected. These observations are clearly critical for understanding how and why globular clusters form, and for using the properties of globular cluster systems to constrain the formation history of their host galaxies.

5

Properties of extragalactic globular cluster systems

Understanding the formation and evolution of galaxies is one of the primary goals of extragalactic astronomy and observational cosmology. The galaxies of the Local Group provide a logical starting point for this effort, and the properties of their globular cluster systems (GCSs) have proved to be a valuable tool in this area, as described in Chapters 3 and 4. In order to gain a broader understanding of galaxy formation, it is necessary to extend these studies to galaxies outside the Local Group. Such studies allow an examination of correlations between galaxies and their globular cluster systems with large sample sizes over a wide range of galaxy properties and environments. Of particular importance are giant elliptical galaxies, which are not present in the Local Group and which, by virtue of their old stellar populations and large masses, are a focal point for tests of galaxy formation models.

Beyond the Local Group, little information can be gleaned about the detailed properties of individual globular clusters, but their integrated properties can be observed to roughly 100 Mpc. In contrast, it is a great technical challenge to resolve stars in galaxies at distances greater than a few Mpc (or several times the size of the Local Group). Thus globular clusters provide discrete dynamical and chemical tracers of galaxies at distances far beyond those where individual stars are observable. The great age of many globular cluster systems is also important, since it implies that these objects are direct probes of the physical conditions at the epoch(s) of galaxy formation. Consequently, globular clusters provide one of the few ways to obtain individual snapshots of the formation history of galaxies.

The focus of this chapter is on the observational properties of extragalactic globular cluster systems. The implications of these observations are the subject of the following chapter. In this chapter, we present the numbers, luminosities, spatial distributions, chemical compositions, and kinematics of the globular cluster systems around galaxies outside the Local Group. For each observational area, the current observational situation is described and the prospects for future advances are discussed.

5.1   Observational requirements

The existence of globular clusters around elliptical galaxies was suggested by Baum (1955) and Sandage (1961), based on long-exposure photographic plates of M87 which showed a large number of unresolved objects clustered around this giant elliptical galaxy. Subsequently, it was shown that the colors of these objects are roughly similar to those of Galactic globular clusters (Racine 1968). These two properties remain the primary observational signature of a globular cluster system: an overdensity of compact objects around a galaxy which have colors in the range expected for old stellar populations.

It is useful to consider in more detail the observational requirements of studies of globular cluster systems of galaxies in the nearby universe. Perhaps the most basic issue is the detection of the globular clusters through imaging. Globular clusters like those in the Galaxy are faint and compact at even the closest distances for which representative samples of galaxies can be studied. For example, at the distance of Virgo, the apparent magnitude of a typical Galactic globular cluster is $V \simeq 24$ and that of the brightest few percent is $V \simeq 21$. The half-light radii of typical Galactic clusters corresponds to roughly 0.1″ at this distance.

Detection of typical globular clusters requires imaging of at least moderate depth. It also benefits greatly from high resolution because of the compactness of the sources. Since the typical magnitudes of these clusters stretch the limits of spectroscopy on 4 m class telescopes, broad-band imaging has been the primary mode of study of significant numbers of objects. This means that individual clusters are rarely confirmed as being associated with the host galaxy on the basis of redshift. Instead, globular cluster systems are identified as a large overdensity of compact objects around the given galaxy.

A consequence of the statistical nature of the detection and study of most extragalactic globular cluster systems is the importance of minimizing contamination from background galaxies and foreground stars. For bright, early-type galaxies, the globular cluster systems are very rich, so that the surface densities can reach roughly 100 clusters per □′near the galaxy centers, and are on the order of tens of clusters per □′at several effective radii of the galaxy. These numbers are significantly in excess of potential contaminating objects at the magnitudes required to study globular cluster populations around galaxies with distances of several thousand km/s or less. However, for detailed analyses of these systems, and for studies of more distant galaxies or those with less abundant cluster systems, minimizing contamination is often a critical factor. Although contamination includes foreground stars, it is dominated at fainter magnitudes by unresolved galaxies. Because nearly all background galaxies are significantly more extended in appearance than globular clusters at the distance of Virgo, high-quality imaging is very valuable for distinguishing between globular clusters and background galaxies. For example, in *HST* imaging around Virgo cluster galaxies, globular clusters are cleanly distinguished from all background galaxies down to the faint magnitude limit of the data (e.g. Whitmore *et al.* 1995). Good ground-based imaging also results in significant improvements in eliminating background galaxies on the basis of image size (e.g. Harris *et al.* 1991).

A third consideration is that the spatial extent of GCSs is usually at least as great as that of the underlying galaxy. Therefore, wide-field imaging is important for building sample size to allow for statistically significant results. Ideally, one would study globular cluster systems with deep, wide-field, high-resolution imaging. Since these are not all necessarily achievable at once, it is interesting to consider how the different obervational factors come into play for typical data that do not reach the peak of the luminosity function. Very roughly, doubling the radius and doubling the sensitivity each give an increase in the number of clusters of about 25%–50%. This

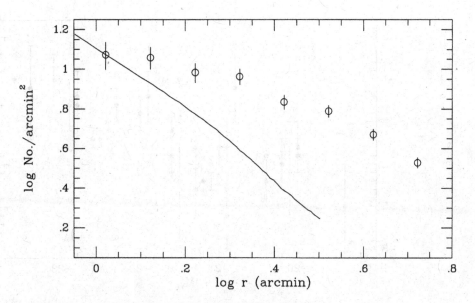

Fig. 5.1. The radial profile of the surface density of globular clusters (dots with error bars) compared to the surface brightness of the integrated light for the elliptical galaxy NGC 4472. (From data provided by D. Geisler.)

approximation breaks down badly at large radii when background contamination becomes significant and at very bright magnitudes at which there are no clusters. Of course, deeper data also give improved precision for brighter objects, and data over a wider field provide more baseline for studies of radial variations of globular cluster properties.

5.2 Spatial distributions

5.2.1 Radial profiles

One of the most general characteristics of globular cluster systems is that they are at least as extended as the light distribution of the galaxy to which they belong. The observational situation is clearest for elliptical galaxies, for which the spatial distribution of the globular clusters is generally more extended than the galaxy light. An example is shown in Figure 5.1. The extended nature of elliptical galaxy GCSs was first suggested by Harris and Racine (1979) and Strom *et al.* (1981), and demonstrated for a larger sample by Harris (1986). All these studies were based on deep photographic plates taken at the prime focus of 4 m class telescopes. Although the plates could not reach the faint magnitudes expected for typical globular clusters, they could reach the brightest fraction of the population. Combined with the wide areal coverage, this was sufficient to delineate spatial profiles for a number of globular cluster systems, particularly the abundant systems of giant ellipticals.

These and subsequent studies with modern CCD detectors have shown that the radial profiles of the GCSs of elliptical galaxies can be well fit either by a power law

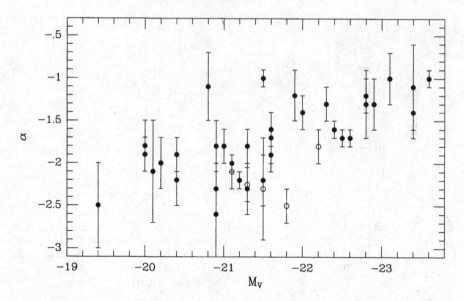

Fig. 5.2. A plot of the slope of the GCS radial profile against galaxy luminosity. The sample is made up of all galaxies for which the slope of the GCS radial profile has been determined (see Appendix 1). Elliptical and S0 galaxies are plotted as filled circles, and spiral galaxies as open circles.

of the form:

$$\Sigma_{GCS} \propto r^{-\alpha}, \tag{5.1}$$

with α ranging from about 1.0 to 2.5, or by a de Vaucouleurs profile,

$$\Sigma_{GCS} \propto \exp[(r/r_e)^{(1/4)} - 1], \tag{5.2}$$

with r_e typically ranging from roughly 10 to 50 h^{-1} kpc. Either of these functions provides a good fit over the whole radial range studied, with the exception of the innermost 1 or 2 kpc. At these small radii, the spatial profile of clusters flattens relative to the best fits to the functions given in equations (5.1) and (5.2) determined at larger radii (Harris 1991 and references therein; also Grillmair *et al.* 1994 and Forbes *et al.* 1996).

The largest source of uncertainty in the determination of the slope of the radial profile is the background level. To some extent this is unavoidable, since the radial profiles are typically determined from imaging data which are limited in radial extent and in which some contamination from foreground and background objects is expected. Studies combining large area with good image quality, like that shown in Figure 5.1, overcome most of these problems and are expected to give reliable results.

The slopes of GCS profiles appear to be correlated with the absolute magnitude of the parent galaxy (Harris 1991). This correlation is shown graphically in Figure 5.2. Since elliptical galaxies themselves tend to have shallower profiles at brighter absolute magnitudes, the *difference* between the profile of the GCS and the galaxy is not obviously dependent on galaxy magnitude. The flattening of the GCS profiles at small radii also appears to correlate with luminosity, with brighter galaxies having larger 'cores' (Forbes *et al.* 1996). Once again, this trend mimics that observed in the galaxy light (e.g. Lauer *et al.* 1995, and references therein). However, it is worth noting that the core radii of the GCS profiles tend to be much larger than those of the underlying galaxy light (Harris 1991; Grillmair *et al.* 1994). Recently, Fleming *et al.* (1995) have suggested that the radial profile of elliptical galaxy GCS may be dependent on environment, with galaxies in clusters having a steeper profile at a given magnitude than galaxies in lower-density regions. More data are needed to test this possibility.

The GCS profiles of spiral galaxies are not as well understood as those of elliptical galaxies. This situation is partly a result of the less populous GCSs of spiral galaxies (see following sections), and partly because it is difficult to disentangle the various luminosity components of the underlying galaxy: the bulge, the disk, and the halo. As discussed in Section 3.6, the radial profile of the halo globular cluster system of the Galaxy is consistent with that of the halo light, with Zinn (1985) finding $r^{-2.5}$ and Harris and Racine (1989) finding r^{-2} for the Galactic GCS, where r is the projected radius. The radial profile of the M31 GCS appears to be similar to that of the Galaxy, with a r^{-2} profile outside the central region (Racine 1991; Fusi Pecci *et al.* 1993a; see also Chapter 4). However, the halo light of M31 appears to be somewhat more concentrated than in the Galaxy (Pritchet and van den Bergh 1994, and references therein).

The situation for other spirals is less well determined, but appears to include galaxies like the Milky Way with similar GCS and halo light profiles (e.g. NGC 4565, Fleming *et al.* 1995) and galaxies which are like M31 in which the GCS is more spatially extended than the halo light (e.g. NGC 7814, Bothun *et al.* 1992). At least some of this variation might be caused by differences between the properties of the halo light of these galaxies, rather than between the properties of the GCSs themselves. Such a variation is commonly accepted as the explanation for the few cD galaxies which have halo profiles that are as shallow as their GCS profile.

Observational constraints on the radial profiles of the GCSs of dE and dSph galaxies are even weaker. Because these systems have few globular clusters per galaxy, it is typical to average the properties of many galaxies together. With this technique, Minniti *et al.* (1996) find that the GCSs of dSphs in the Local Group appear to be more extended than the galaxy light, consistent with the earlier work of Harris (1986). Durrell *et al.* (1996) employed a similar technique in their study of the GCS profiles and found that the average GCS profile is consistent with the average integrated light profile. They also showed that individual GCS profiles appear to vary, with some the same as the galaxy light, and others more extended. The small number of clusters for these dE galaxies translates into large uncertainties, however.

The observations lead to a general picture in which the spatial profiles of GCSs are similar to or more extended than the spheroidal component of their host galaxy. Like the spheroidal components, the spatial profiles of GCSs are more extended for galaxies which are brighter and of earlier type. However, in the case of GCSs, it is currently difficult to determine whether it is galaxy morphology or luminosity that is the dominant factor in determining the GCS profile, primarily because of the small number of spirals for which suitable GCS data are available.

5.2.2   *Azimuthal distributions*

Considerably less is known about the shapes of GCSs than about their radial profiles. For the most part, this is the result of the difficulty of determining the ellipticity and position angle (PA) of a distribution with the modest number of points usually available (100–200 is typical). Moreover, the region sampled by these points usually has a complex geometry with an outer cutoff set by the size of the CCD and an inner cutoff caused by the high surface brightness of the galaxy near its center. The inner cutoff is particularly problematic because its shape is that of the galaxy light, and because many clusters are near it.

Despite these difficulties, there are some results on the shapes of GCSs of elliptical galaxies. The available data are all consistent with a close agreement between the position angles of the GCS and the elliptical galaxy light. This result has probably been established most clearly for M87 (McLaughlin *et al.* 1994, Cohen 1988), and NGC 720 (Kissler-Patig *et al.* 1996), for which $\Delta PA \lesssim 10 - 20°$. Similarly, Forbes *et al.* (1996) find no significant offset between GCS and galaxy light position angles when they average over a sample of 14 ellipticals with fewer detected globular clusters. Although these studies have demonstrated that there is at least a rough similarity in position angles, it is not yet clear how exact this agreement is and whether it applies to all elliptical galaxies.

The ellipticities of the GCSs of elliptical galaxies also appear to be roughly correlated with those of the host galaxies, but more detailed comparisons do not yet give a clear picture. Examples of published data include the studies of the GCSs of M87 and NGC 720 listed above. Both of these show that the GCS ellipticity is not very different from that of the integrated galaxy light, but are consistent with a GCS ellipticity ranging from somewhat more to somewhat less flattened than the galaxy light. For galaxies of slightly later type, there is some evidence that the GCS is rounder than the integrated light of the spheroidal component (e.g. NGC 3115, Hanes and Harris 1986, and NGC 4594, Harris *et al.* 1984). However, this may not be universally true of galaxies with disks, as the GCS of the spiral galaxy NGC 7814 may be highly flattened (Bothun *et al.* 1992).

5.3   Luminosity distribution

A basic observable quantity of globular cluster systems is the number of objects per luminosity or magnitude interval, traditionally called the luminosity function. The luminosity function of the Galactic globular cluster system was discussed in Section 3.2, and that of M31 in Section 4.1. It is natural to attempt to extend the analysis of globular cluster luminosity functions (GCLFs) to galaxies beyond the Local Group. Much of this work has been motivated by the hope of using the turnover of the GCLF as a distance indicator. Understanding the GCLF is also important for calculating the total number of clusters, since the entire population is rarely observed. Moreover, the GCLF provides constraints on the dynamical evolution of the globular cluster population (see Section 3.5) and on theories of globular cluster formation (see Chapter 7).

When considered on the logarithmic magnitude scale, the $N(M)$ of the GCLF of the Galaxy and M31 is closely approximated by a Gaussian distribution of the form

$$N(M) = N_0 e^{-(M-M_0)^2/2\sigma^2}, \tag{5.3}$$

with a mean magnitude for the two systems of $M_{0,V} = -7.4 \pm 0.2$ and a dispersion of $\sigma = 1.2 \pm 0.1$. As we discussed in Chapters 3 and 4, the magnitudes of the GCLF peaks in the Milky Way and M31 are consistent with one another. The Harris *et al.* (1991) distance calibration for Milky Way globular clusters leads to a slightly fainter peak in the Milky Way compared to M31, but the difference is not statistically significant. The uncertainty on the mean GCLF peak of the two galaxies quoted above encompasses current estimates of this quantity for both the Milky Way and M31 systems. Because of this close agreement between the well-studied GCLFs of these Local Group spirals, their mean values provide a useful benchmark against which the peaks and dispersions of the GCLFs of other galaxies can be compared.

Ideally, one would observe all the globular clusters around a given galaxy, and compare the resulting $N(m)$ directly with that observed for the Galaxy and M31. Such a measurement would immediately answer many questions about the formation and dynamical evolution of globular cluster populations, as well as testing the universality of the turnover of the GCLF and its accuracy as a distance indicator. Unfortunately, the faintness of typical globular clusters at distances of interest (e.g. galaxies in the Virgo cluster) has precluded attempts at such a direct comparison. The deepest exposures to date on 4 m class telescopes have only reached slightly past the turnover for the populous GCSs of Virgo ellipticals (e.g. Harris *et al.* 1991), and attempts to push deeper from the ground will require significant advances in distinguishing globular cluster from background galaxies at very faint magnitudes. This situation is beginning to change with imaging utilizing the refurbished *HST*. However, these data are not yet common, so it is worthwhile examining what can be gleaned from more typical ground-based studies.

5.3.1 GCLFs from ground-based imaging

An example of an observed GCLF for an elliptical galaxy is shown in Figure 5.3. This example does not represent the limit of current ground-based imaging, which is somewhat deeper, but is typical of the majority of the available data. Observations like these clearly do not allow a comparison which is direct and free of assumptions to the GCLFs of the Milky Way and M31. Instead, the data are usually fit by a function, most often the Gaussian described earlier. Hanes and Whittaker (1987) showed that all three parameters of the Gaussian description (N, m_0, and σ) cannot be reliably constrained with data like those shown in Figure 5.3, which do not go fainter than the turnover of the GCLF. Secker and Harris (1993) expanded on this work and showed that it applies to other parametric descriptions of the GCLF and that unconstrained fits to data which do not reach well past the turnover lead to biases in the fitted values.

Although an independent determination of the three parameters characterizing the Gaussian description of the GCLF requires data which reach significantly past the turnover, some progress can be made on more limited data given additional assumptions. For example, Hanes and Whittaker (1987) showed that if one of the parameters of the Gaussian form is fixed, then data like the NGC 3923 GCLF shown in Figure 5.3 can provide good constraints on the other two parameters, if the adopted form of the GCLF is assumed to be correct. For the specific case of NGC 3923, if m_0

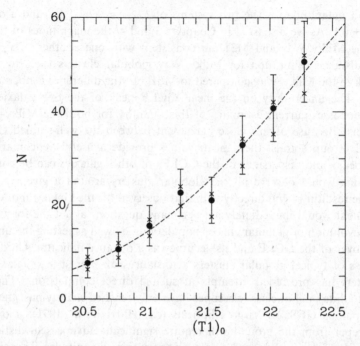

Fig. 5.3. A plot of the GCLF of NGC 3923 from Zepf *et al.* (1994). The filled circles represent the number of clusters using the standard background values, with the crosses representing the upper and lower limits for the background. The dashed line is the fit to these data with $m_0 = 23.8, \sigma = 1.51$. As discussed below, other m_0, σ combinations provide an equally good fit within the magnitude range of the data. The T_1 bandpass is similar to R_C.

is fixed to be 23.8, then $\sigma = 1.51 \pm 0.10$, with statistically indistinguishable fits being obtained for a range of m_0, σ combinations (e.g. Zepf *et al.* 1994).

Since it is sometimes possible to obtain relative distances between galaxies which are thought, or at least hoped, to be reliable, the GCLFs of various galaxies can be compared to test for the universality of M_0 and σ. This is often done for ellipticals in the cores of nearby clusters (e.g. Harris *et al.* 1991, Secker and Harris 1993, and Ajhar *et al.* 1994 in the Virgo cluster; Blakeslee and Tonry 1996 and Kohle *et al.* 1996 in the Fornax cluster). These studies have shown that the GCLF turnover appears to vary by less than about 0.2 magnitudes among elliptical galaxies within each of these clusters (see also Whitmore 1997). This conclusion is somewhat dependent on the assumption that σ is constant among the GCLFs of these galaxies, although it would seem improbable for σ to vary in such a way as to make different intrinsic values of M_0 appear identical. Similarly, the constancy of M_0 for the GCSs of ellipticals in the Virgo and Fornax clusters is directly dependent on the assumed relative distances between the ellipticals within each cluster. However, for most of the galaxies, the simple assumption that objects in a given cluster core are at the same distance is all that is required.

The deepest ground-based data reach sufficiently past the GCLF turnover that some constraint on the dispersion, σ, is possible without fixing m_0 (e.g. Harris *et al.*

1991). These data suggest that the GCLFs of giant ellipticals are best characterized by $\sigma \simeq 1.4$ rather than the value of 1.2 found for the Milky Way and M31. Similarly, better fits for $\sigma \simeq 1.4$ are found for other ellipticals (e.g. NGC 1399 (Bridges *et al.* 1991) and NGC 5128 (Harris *et al.* 1984)), *when* m_0 is fixed by other considerations. However, larger σ values for elliptical galaxy GCLFs may not be universal, as Kissler-Patig *et al.* (1996) find $\sigma \simeq 1.2$ for NGC 720. It is important to keep in mind that the data discussed here only marginally break the m_0, σ degeneracy, and that conclusions regarding σ may also be affected by the Gaussian fit to the observed GCLF. These issues are directly addressed by *HST* imaging discussed in the following subsection.

Because data like those of Harris *et al.* (1991) are sufficiently deep that the estimates of m_0 and σ are at least partially decoupled, it is also possible to compare the turnover luminosities estimated from these data with those found for the GCLFs of the Milky Way and M31. This comparison indicates that either $M_{0,V}$ is typically about 0.2–0.3 magnitudes fainter for Virgo ellipticals than Local Group spirals or that there is an error of about 10–15% in the distance modulus to Virgo of 31.0 derived from secondary distance indicators (Jacoby *et al.* 1992). A similar discrepancy is observed for the Fornax samples of Blakeslee and Tonry (1996) and Kohle *et al.* (1996), as well as other ellipticals, given $\sigma \simeq 1.4$, as appears to be observed. If the RR Lyrae calibration advocated by Sandage and collaborators is used to determine distances to Milky Way globular clusters (see Chapter 3), the discrepancy between the GCLF and other secondary distance indicators is increased by an additional ~ 0.2 magnitudes.

A possible explanation for the small differences between the M_0 values of spirals and ellipticals was given by Ashman *et al.* (1995). They pointed out that even if the globular cluster systems in ellipticals and spirals have the same *mass function*, they will have slightly different *luminosity functions*, since elliptical galaxy GCSs are slightly redder on average than spiral galaxy GCSs, probably because of metallicity differences (see Section 5.5 below). Because mass is expected to be the physical quantity of interest, but luminosity is observed, it is useful to account for this effect. Using stellar populations models to give luminosity corrections in a number of bandpasses as a function of metallicity (color) differences, Ashman *et al.* (1995) found that for globular clusters of equal mass, those around the Milky Way and M31 are expected to be about 0.2 magnitudes brighter in V than typical clusters around ellipticals because of the bluer colors of the Milky Way GCS. This correction generally works even if the difference in color is assumed to be caused partially by age differences as well as by metallicity differences (Zepf 1996).

An alternative explanation of the modest variation of M_0 found by a comparing the GCLF to other distance indicators is that M_0 is brighter for galaxies in low-density environments (Blakeslee and Tonry 1996). Since spiral galaxies occupy lower-density regions and have bluer globular cluster systems than ellipticals, it is clear the color effect described above provides a natural mechanism for any correlation with local density. However, Blakeslee and Tonry (1996) find that the difference is greater, although with large uncertainties, than can be accounted for by the color corrections of Ashman *et al.* (1995). Much more accurate data are required to conclusively determine whether the color-corrected value of M_0 is independent of environment. This is particularly true given evidence that σ is not constant among elliptical galaxy GCLFs (see the discussion above and in the following section).

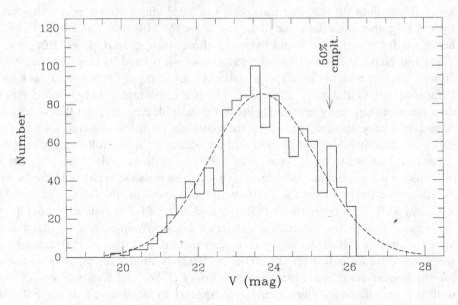

Fig. 5.4. A plot of the M87 GCLF from the Whitmore *et al.* (1995) *HST* data. Their best-fitting Gaussian is shown in the dashed line, and the 50% completeness limit is marked.

It has also been suggested that the GCLF peak in dwarf galaxies is fainter than in giant galaxies (Harris 1996). A definitive result is not yet possible given the small number of globular clusters in each system, and the modest number of systems as a whole, but the topic clearly warrants further study. It is notable that any such effect cannot be accounted for by color differences because the GCSs of dwarf galaxies and spiral galaxies have similar colors. Possible mechanisms for producing differences in the peak and the dispersion of the GCLF are discussed in the following chapters.

5.3.2   GCLFs from HST imaging

Data from the *HST* are now available which reach magnitudes significantly fainter than the GCLF turnover. The great advantage of these data is that they go much fainter than the turnover with a large sample and minimal contamination. As a result, the determinations of m_0 and σ are mostly decoupled from one another. The Whitmore *et al.* (1995) study of the M87 GCS utilizing *HST + WFPC2* (Wide-Field/Planetary Camera 2) was the first to clearly define the GCLF much fainter than the turnover for a galaxy at this distance. Their data and best fit to the GCLF is shown in Figure 5.4. This plot demonstrates dramatically that at least this elliptical galaxy has a GCLF of the same general form as those of the Local Group spiral galaxies, the Milky Way and M31. Fitting these M87 GCLF data to a Gaussian form, Whitmore *et al.* (1995) find that $m_{0,V} = 23.72 \pm 0.06$, and $\sigma = 1.40 \pm 0.06$.

Given the observed $m_{0,V}$ and a distance modulus of 31.04 ± 0.17 for the Virgo cluster derived from Cepheid observations of the galaxy M100 (Ferrarese *et al.* 1996; Freedman *et al.* 1994), the resulting $M_{0,V}$ for the M87 GCS is -7.3 ± 0.2, where the uncertainty only reflects the formal uncertainties of the fit to the M87 GCLF and the

Cepheid-based distance modulus to M100. The corresponding average value for the Milky Way and M31 is -7.4 ± 0.2 (see Sections 3.2.1 and 4.1.1 for a discussion of this value and its uncertainties). These data are therefore consistent with a universal GCLF. They are also equally consistent with a universal mass function for globular clusters and a correction for color differences between the M87 GCS and the local calibrators which causes the M87 GCLF to be shifted fainter by about 0.2 magnitudes in V relative to the GCLF of the Milky Way and M31 (Ashman *et al.* 1995). These results are slightly inconsistent with the local calibration of $M_{0,V}$ advocated by Sandage and collaborators, unless the color corrections are also adopted. The overall conclusion is that the GCLF, and its underlying mass function, appear to be very similar in M87 and Local Group spirals, with no detectable differences, and a 1σ uncertainty in the comparison of about 0.3 magnitudes. This regularity is striking given the vast differences between the parent galaxies of the globular clusters, as well as the differences in number and color between the GCSs themselves.

Because the *HST + WFPC2* images of M87 are clearly reliable at magnitudes fainter than the turnover, the derived value of the GCLF dispersion of $\sigma = 1.40$ is well determined. This lends support to earlier suggestions that the elliptical galaxy GCSs have luminosity functions which are slightly broader than those of spirals (Harris 1991). However, the dispersion of the M87 GCLF is not quite as broad as suggested by McLaughlin *et al.* (1994) and Blakeslee and Tonry (1995), indicating the importance of the *HST* data for addressing these questions.

Although the Whitmore *et al.* (1995) data conclusively show that the σ of the M87 GCLF is broader than those of Local Group spirals, *HST* imaging studies of two other ellipticals indicate that this effect is not universal. Specifically, Forbes (1996) finds that the GCLFs of NGC 4278 and NGC 4494 are well fit by Gaussians with $\sigma \simeq 1.1$, similar to the Milky Way and M31, but less than those derived for giant ellipticals in Virgo and Fornax. This result confirms the earlier suggestion by Kissler-Patig *et al.* (1996) that not all elliptical galaxy GCLFs are equally broad. Variations in σ for elliptical galaxy GCLFs have important implications for comparisons of M_0 based on data that do not reach well past the turnover magnitude. Similarly, they may prove to be the limiting factor for accurate determinations of m_0 for GCLFs at distances for which an independent determination of σ is not possible, such as the *HST* observations of NGC 4881, an elliptical galaxy in the Coma cluster (Baum *et al.* 1995).

It appears very likely that *HST* imaging will play a major role in advancing our understanding of the GCLF. For example, constraints on radial variations in the GCLF are fairly limited outside the Local Group. The best example from ground-based studies is probably the M87 GCS, for which McLaughlin *et al.* (1994) find no variation with radius (see Figure 2 of McLaughlin and Pudritz 1996). Because *HST* images can reach well past the turnover, they can place strong constraints on any radial variation in the turnover by imaging at several positions in a given galaxy. Such variations are an important probe of dynamical destruction in GCSs (see Chapter 7).

A second example of an area which will be advanced by upcoming *HST* studies is the use of the GCLF turnover as a distance indicator. The recent results on the M87 GCS (Whitmore *et al.* 1995) are promising, although there are still several remaining concerns about possible systematic uncertainties in the luminosity of the

GCLF turnover. Chief among these is constraining any possible differences between the GCLF turnover in elliptical galaxies, which are the most promising targets for distance scale programs, and those in the Milky Way and M31 which currently provide the calibration of the turnover luminosity.

As described earlier, the M87 data suggest that the difference between the GCLF turnover luminosity in ellipticals and spirals is less than 0.2 magnitudes. This comparison is based on Cepheid distances to spirals believed to be in the Virgo cluster, and accounts for the known color difference. Similarly good agreement is found by comparing less well-determined GCLF turnover luminosities in other ellipticals with secondary distance estimates for these galaxies (Ashman *et al.* 1995). These results suggest that the GCLF may have a promising future as a distance indicator, although it will clearly benefit from further tests of its reliability. One good way to test the similarity of the GCLF turnover in ellipticals and spirals is to observe ellipticals which are located in small, well-defined groups which also contain spirals with a Cepheid distance. The GCLF turnover in spiral galaxies can also be more tightly constrained by using *HST* to observe globular clusters in spiral galaxies which have Cepheid distance determinations.

5.4   Total number of clusters

Observations of the radial profile and luminosity function of extragalactic globular cluster systems can be combined to give the total number of clusters (hereafter N_{GC}) in any given galactic system. Such a calculation is of natural interest for understanding the efficiency of globular cluster formation (and perhaps destruction as well). It was already apparent from Sandage and Baum's deep exposure plates of M87 that this giant elliptical galaxy had a much richer globular cluster system than that of the Galaxy. Further work with photographic plates by Hanes, Harris, and others (e.g. Hanes 1977) extended these estimates to a larger number of galaxies, with a particular emphasis on Virgo ellipticals (see reviews by Harris 1991 and Harris and Racine 1979). Subsequent studies using CCD detectors allowed deeper measurements, but have only recently contributed significantly to determinations of N_{GC} because of the limited areal coverage of most CCDs before the large arrays of the 1990s.

One of the first conclusions to emerge from these early studies is that elliptical galaxies have more globular clusters per unit luminosity than spiral galaxies. A common way to normalize the number of clusters by the luminosity of the parent galaxy is the specific frequency S_N, which is defined as:

$$S_N \equiv N_{GC} 10^{-0.4(M_V+15)} \tag{5.4}$$

(Harris and van den Bergh 1981). This normalization has the advantage that it is observationally straightforward. However, ellipticals and spirals have different stellar populations and therefore different V-band mass-to-light ratios, so a refinement to the specific frequency is required to determine how the number of globular clusters varies as a function of stellar *mass* for various galaxy types. One approach is to change the normalization of S_N to include only the spheroidal component of a galaxy, for which the mass-to-light ratio is less likely to vary for different galaxy types (Harris 1991). However, as Harris noted, this approach is limited by the difficulty of determining the luminosity of the spheroidal component for spiral galaxies, especially those of very

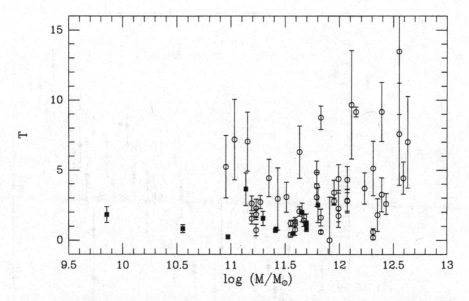

Fig. 5.5. The mass-normalized specific frequency T plotted against estimated galaxy mass. Early-type galaxies are represented by open circles and spiral galaxies by filled squares (data from Appendix 1).

late type, like the LMC. Moreover, in the Milky Way, the halo globular clusters do not appear to be physically associated with the central bulge.

An alternative approach was introduced by Zepf and Ashman (1993), who normalized N_{GC} directly by the estimated stellar mass of the host galaxy, M_G, by using a parameter, T, defined as

$$T \equiv \frac{N_{GC}}{M_G/10^9 \, M_\odot}. \tag{5.5}$$

The weakness of this approach is that it requires an additional quantity: the stellar mass-to-light ratio of various galaxies. Zepf and Ashman (1993) used mean values for the various galaxy types determined by spectroscopic studies (e.g. Faber and Gallagher 1979). These relative stellar mass-to-light ratios are also consistent with those derived from the application of stellar population models to the observed colors of different galaxy types. Thus, for comparisons between samples of galaxies, for which only the relative values matter, the T parameter is likely to be reliable. A compilation of available T values for spirals and ellipticals is shown in Figure 5.5. This figure shows that ellipticals have T values which are about a factor of two or three higher than those of spirals, although both the statistical and systematic uncertainties are still large.

Figure 5.5 also shows that even among early-type galaxies, there are variations in the number of globular clusters per unit luminosity. Specifically, there appears to be a general trend of increasing T (or S_N) with elliptical galaxy luminosity, along with significant scatter at any given luminosity. In order to study further how N_{GC} for early-type galaxies varies with other parameters, we plot the number of globular

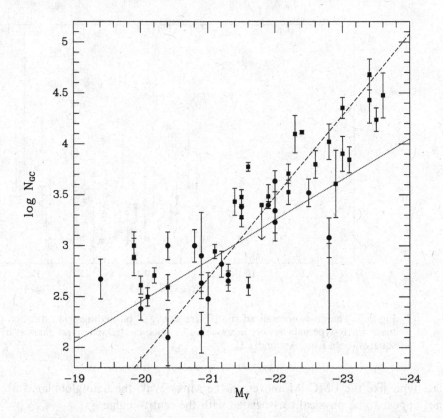

Fig. 5.6. Log of the total number of globular clusters, N_{GC}, plotted against the absolute magnitude of the host early-type galaxy. Galaxies in clusters and rich groups ($\sigma > 400$ km/s) are plotted as squares, and galaxies in poorer environments ($\sigma < 400$ km/s) are plotted as circles. The dashed line indicates the slope for $N_{GC} \propto L^2$, and the dotted line corresponds to $N_{GC} \propto L^1$ (data from Appendix 1).

clusters against the absolute magnitude of the host elliptical galaxy in Figure 5.6. Although the error bars are large and there appears to be real scatter, this plot shows that the slope of the relation between the number of globular clusters and absolute magnitude is greater than one, but less than two. Thus, the number of globular cluster per unit luminosity (S_N) is larger for bright ellipticals than for faint ones (Djorgovski and Santiago 1992; Zepf *et al.* 1994). If a single relation between N_{GC} and L holds for early-type galaxies, then this can be expressed as $N_{GC} \propto L^\alpha$, with $\alpha > 1$. Alternatively, the trend may be described as an offset to higher S_N values for more luminous ellipticals compared to lower luminosity ellipticals.

Historically, it has been difficult to disentangle the increase of S_N with luminosity from an environmental effect, since most of the bright ellipticals which had been observed were in clusters, and most of the faint ones were in the field. As a result, some of the elliptical galaxy data can also be described by a correlation of S_N with local density (e.g. Kumai *et al.* 1993a; West 1993), although the quality of the fit is dependent on the definition of local density. The first evidence that luminosity is the primary parameter determining S_N came from Djorgovski and Santiago (1992), who used a multivariate analysis of the data then available. This conclusion is supported by the discoveries of Zepf *et al.* (1994) and Hopp *et al.* (1995) of bright ellipticals outside of clusters with rich globular cluster systems. Thus, it appears that the increase of S_N with L is the primary factor in determining N_{GC}, although other secondary effects are not excluded and, in fact, appear to be necessary given the significant scatter at all luminosities seen in Figure 5.6.

The scatter about any relationship between the number of globular clusters and properties of the host galaxy has been best studied for very bright ellipticals and cD galaxies. Some of these, such as M87 and NGC 1399, have remarkably rich GCSs, but other, very similar, galaxies have many fewer globular clusters. These differences in S_N have little correlation with other properties of either the host galaxy, or the cluster or group in which it resides (e.g. Harris *et al.* 1995; Bridges *et al.* 1996a). West *et al.* (1995) claim that a correlation exists with cluster properties if they are normalized by a factor accounting for the distance of the galaxy from the center of the cluster relative to the core radius of the cluster. The core radius is assumed to be the same for all clusters. However, the correlation seems to disappear if the core radius increases with cluster temperature.

Studies of the relationship between N_{GC} and L for elliptical galaxies have also been extended to dwarf ellipticals by Durrell *et al.* (1996a). They find that the slope which fits the elliptical galaxy data best ($1 < \alpha < 2$) underpredicts the number of globular clusters around dwarf ellipticals if extended to these faint galaxies. Moreover, for dwarf ellipticals, the number of globular clusters per unit luminosity appears to *decrease* with increasing luminosity, exactly opposite to the trend for normal ellipticals. This may be related to similar trends in mass-to-light ratios, which increase with luminosity for normal ellipticals, and decrease with luminosity for dwarf ellipticals.

Before venturing further, it is worthwhile considering the limitations of the determinations of N_{GC} described above. The most sobering aspect of this consideration is the small fraction of the total number of clusters which is actually observed for a typical galaxy measurement. For readily available cameras and CCD detectors, about one-third of the total number of estimated clusters typically falls within the field of view. For typical exposure times and object distances, the sample becomes seriously incomplete at about one magnitude brighter than the turnover of the GCLF, and thus encompasses about one-third of the total estimated number of clusters within the area observed. Combining these two implies that total cluster estimates are based on observation of only about 10% of the putative population! It is important to emphasize that modern observations are capable of doing much better than this, but the description given above applies to much of the data that one must currently use to analyze large samples.

This analysis indicates that most current determinations of the total number of clusters are dependent on extrapolations of the radial profile and luminosity function, and thus should be considered warily. However, for some purposes, the situation is significantly better. For example, the same parts of the radial profile and luminosity function are studied for typical elliptical galaxies in the sample, so galaxy-to-galaxy comparisons are fairly robust, even if the total estimates are less reliable. For this reason, calculations such as the increase of S_N with L described above are likely to be correct.

One way to avoid some of the extrapolations required when dealing with the globular cluster systems and the galaxies as a whole is to consider a localized specific frequency of globular clusters, often written as $S_N(r)$. This quantity has been found to increase steadily with galactocentric radius, with typical values in the regions observed which are slightly less than the estimated global values. (e.g. McLaughlin *et al.* 1994). Such behavior is expected given the more extended radial profile of the GCS relative to the elliptical galaxy light (Section 5.2.1).

A second quantity which may be less subject to some of these uncertainties for typical extragalactic GCSs is the total luminosity/mass of the GCS. This is because most of the total luminosity of a GCS is contained in the most luminous clusters which are more readily observed (Harris 1991). Regardless of the exact technique used, galaxy-to-galaxy comparisons indicate that within the luminosity and radial range typically observed, ellipticals generally have richer globular cluster systems than spirals and that brighter ellipticals have more clusters per unit luminosity than fainter ellipticals. Elucidating trends between the number of clusters and galaxy type and luminosity require more observations, particularly of spiral galaxies and low-luminosity ellipticals.

5.5   Average color

Colors have long played a major role in the study of extragalactic globular cluster systems. Much of this importance stems from the fact that (broad-band) colors reflect the age and metallicity of the stellar population (see Chapter 2). Early in the study of these systems, the identification of bright, compact objects seen around galaxies like M87 as globular clusters was supported by the rough similarity of their colors to Galactic globular clusters (Racine 1968). Subsequently, a photographic study concentrated on the colors of the GCSs of several Virgo ellipticals (Strom *et al.* 1981; Forte *et al.* 1981). Two important results emerged from this work. The first was the claim that elliptical galaxy GCSs have a color gradient such that clusters become bluer on average with increasing galactocentric distance. This initially contentious result has been confirmed by subsequent studies, as we describe in the next section. The second finding, again confirmed by later studies, is that the globular clusters are bluer on average at a given radius than the integrated light of the elliptical galaxy at that same radius. Generally, the average cluster color at the effective radius of the galaxy light is about 0.1 bluer in $(B - V)$ than the integrated elliptical galaxy light at the same radius (e.g. Harris 1991). However, color gradients in the GCS and the integrated galaxy light (and the differences between these two) make it difficult to establish a more precise characteristic value for the color difference. It is also notable that the color distribution within an elliptical galaxy GCS is broad at all radii (Section 5.7.1).

The comparison between the average colors of different GCSs is a powerful tool for constraining the formation histories of different galaxy types. Such comparisons have now established that the GCSs of elliptical galaxies have significantly redder colors than those of spiral galaxies like the Milky Way and M31. This result was first clearly demonstrated by Cohen (1988), who showed that the GCSs of several bright

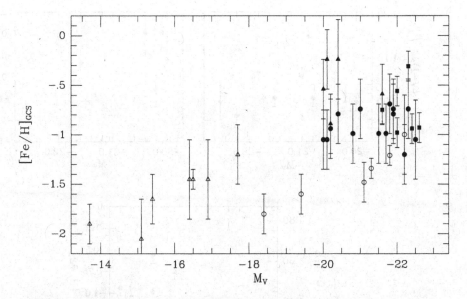

Fig. 5.7. A plot of the estimated metallicity of the GCS against the absolute V magnitude of the host galaxy. Elliptical galaxies are plotted as solid symbols, spiral galaxies as open circles, and dSph and dE galaxies as open triangles. These data show that elliptical galaxies have more metal-rich GCSs than spirals or dwarf ellipticals. There is no clear trend of metallicity with host galaxy luminosity, except that driven by the correlation in this sample between galaxy type and luminosity. Note that the elliptical galaxies are drawn from three samples, each coded with a different symbol. Solid squares represent metallicity estimated from $(C - T1)_0$, solid triangles ground-based $(V - I)_0$ data for Fornax ellipticals, and solid circles $(V - I)_0$ data from *HST* studies. The error bars include the uncertainty in the location of the mean or median of the color and in the foreground reddening, and also attempt to account for the uncertainty associated with converting color to metallicity. All the data in this plot are listed in tabular form in Appendix 1.

Virgo ellipticals are redder than the Galactic GCS. This area of research has been particularly active in the mid-1990s, as large format CCDs have made it possible to obtain accurate photometry for large samples of globular clusters in a number of galaxies.

A compilation of all the available colors of GCSs is plotted against the absolute magnitude of the host galaxy in Figure 5.7, with different morphological types noted by different symbols, as described in the figure caption. A wide variety of colors and photometric systems have been used in these surveys, so a first order attempt was made to bring them onto a similar system by using the color–metallicity calibration for Galactic globular clusters in each system.

Although Figure 5.7 is useful for giving a sense of the trends of GCS color with morphology and luminosity, for a more detailed comparison it is useful to restrict the samples to data taken in the same way on the same photometric system. Three useful samples of elliptical galaxy colors which meet these criteria are a shallow *HST* survey of elliptical galaxy GCSs in *V* and *I* (Forbes *et al.* 1996), a ground-based *V*,*I* survey

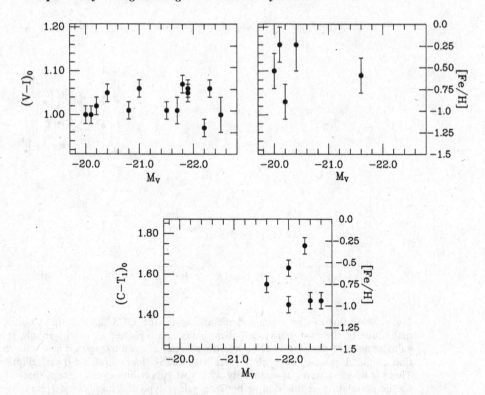

Fig. 5.8. The average GCS color is plotted against the absolute magnitude of the host galaxy for three surveys. The upper left is from Forbes et al. (1996), the upper right from Kissler-Patig *et al.* (1997), and the lower plot is from Zepf *et al.* (1995), Geisler *et al.* (1996), and references therein. The error bars include uncertainties in the photometry and foreground reddening, but not in the conversion of colors to metallicity.

of GCSs of early-type galaxies in the Fornax cluster (Kissler-Patig *et al.* 1997) and a compilation of ground-based results using the $(C - T_1)$ color of the Washington photometric system (Zepf *et al.* 1995a; Geisler *et al.* 1996, and references therein). In Figure 5.8 we plot the GCS color against host galaxy magnitude for each of these three surveys independently.

The formal uncertainty in the median or mean colors plotted in these figures is small, primarily because of the large number of clusters ($\gtrsim 100$). While the color distributions are non-Gaussian, robust statistical estimators (which make no assumption about the form of the parent distribution) show that the uncertainty in the mean is close to the familiar $\sigma/\sqrt{N}$ for a Gaussian distribution. Given the small statistical error, systematic uncertainties arising from the use of different photometric systems and corrections for reddening along the line of sight may be significant. The former can be dealt with by only considering data taken in the same photometric system (as is done in Figure 5.8). The latter is fortunately smaller than the observed differences between GCSs, as shown in the figures. Finally, because radial color gradients appear to be a common feature of GCSs (as described in the following section), the mean or median color will depend slightly on the radial range covered.

In practice, the gradients are not steep and most systems are observed over similar radial ranges, so this is a small or negligible effect.

Figures 5.7 and 5.8 clearly demonstrate that elliptical galaxy GCSs have redder colors than spiral galaxy GCSs. The implications of this result are discussed in the following chapter. The figures can also be interpreted as a trend of increasing GCS color with increasing galaxy luminosity (van den Bergh 1975; Brodie and Huchra 1991). The few globular clusters of faint dwarf ellipticals and dwarf spheroidals are roughly consistent with this trend, as they have average colors which are as blue or bluer than those of spiral galaxies. Because luminosity and morphology are tightly coupled in the current samples, it is difficult to disentangle the influence of these host galaxy parameters on GCS color. However, the current data appear to indicate that the GCSs of spirals have lower metallicity than the GCSs of ellipticals of the same host galaxy magnitude

A second way to attempt to decouple the effects of luminosity and morhphology is to consider subsamples composed of only one morphological type. Figure 5.8 shows that the average colors of the elliptical galaxy GCSs are all roughly similar, with little or no trend with luminosity or other galaxy property. Rather than any strong trend, Figure 5.8 shows significant scatter in the average color for elliptical galaxy GCSs. This scatter appears to be larger than the individual uncertainties, particularly for $(C - T1)_0$ data, which are the most metallicity sensitive. As discussed above, the samples from which this conclusion is drawn should be largely free of spurious galaxy-to-galaxy variations. Thus, there appears to be real scatter among the colors of elliptical galaxy GCSs.

The variations in the color of elliptical galaxy GCSs do not appear to correlate with any properties of either the host galaxy or the GCS. In Figure 5.9, we plot S_N against GCS metallicity, using the same samples of spirals and ellipticals as were used in Figure 5.7. This figure shows again that giant ellipticals have both higher S_N values and GCSs of higher average metallicity than spiral galaxies. Figure 5.9 also reveals that among elliptical galaxies there is no obvious correlation of average GCS color with S_N. These samples are not large, and further work may yet reveal some correlations, but at the current time the average GCS color has little correlation with any other parameter.

Throughout this section, we have focussed on the colors of globular cluster systems. In the literature, broad-band colors of extragalactic GCSs are often immediately translated into metallicities, given a color–metallicity calibration, either based on the Galactic GCS (e.g. Couture *et al.* 1990; Geisler and Forte 1990) or on theoretical models (e.g. Worthey 1994). The assumption that GCS colors primarily reflect metallicity is likely to be sound. From detailed studies of individual Galactic globular clusters, we know these objects are old and relatively metal-poor (see Chapter 2). Moreover, Galactic globular clusters with intrinsically redder colors are redder primarily because of their higher metallicity (see Chapter 3). Thus, the most straightforward explanation for the redder colors of elliptical galaxy GCSs compared to spiral galaxy GCSs is that globular clusters around elliptical galaxies have higher metallicities on average. The difference between the average colors of elliptical and spiral galaxy GCSs shown in Figure 5.7 corresponds to a metallicity difference of about 1.0 dex in [Fe/H]. If the redder colors of elliptical galaxy GCSs relative to the Milky Way GCS are ascribed

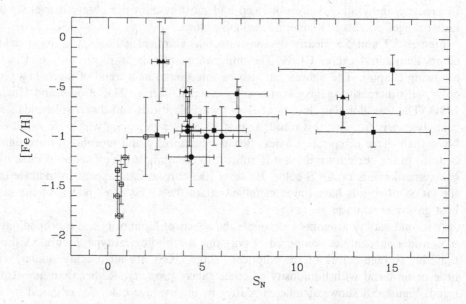

Fig. 5.9. The S_N values of ellipticals and spirals plotted against the average [Fe/H] of the GCS. The samples and symbols are the same as in Figure 5.7. Ellipticals galaxies have richer and more metal-rich GCSs than spirals, but within the sample of ellipticals, no correlation of average [Fe/H] with S_N is observed.

solely to age rather than metallicity, their ages must be extraordinarily large. This simply reflects the insensitivity of broad-band colors to age once a stellar population is older than a few Gyr, and the sensitivity of these same colors to metallicity differences (see Section 2.1).

Although broad-band colors primarily reflect metallicity, they may not do so exclusively. In detail, globular clusters are probably not coeval, either within a given GCS or between GCSs. We already know that there is an age spread within the Galactic GCS. Moreover, the best calibration between specific broad-band colors and abundance is likely to change, and improve, with time. A further complication is that the ratios of elemental abundances may differ between extragalactic systems and the Galactic clusters which serve as the basis for color–metallicity calibrations. For these reasons, we prefer to give the results in terms of the observed colors, and provide the current best estimates of how these translate into metallicities. We have done this when possible. However, when comparing samples obtained in different color systems, the Galactic color–metallicity calibration provides a way to attempt to put these on the same scale.

5.6   Color gradients

Color gradients of GCSs are of natural interest for the constraints they place on the formation history of the system. They are particularly valuable for studying enrichment and dissipation in the formation process. As an example, the Galactic GCS has an overall color gradient that is driven by the smaller galactocentric radii of the thick disk clusters, which have higher metallicity than the more extended halo

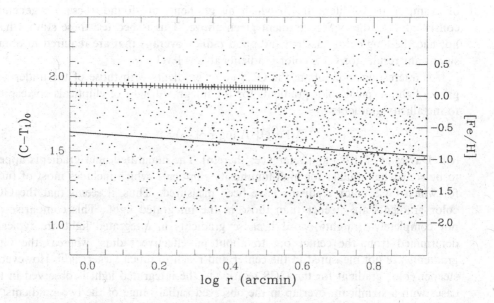

Fig. 5.10. $(C - T_1)_0$ vs. radius for NGC 4472 with globular clusters plotted as points and the integrated light as plus signs (adapted from Geisler *et al.* 1996).

cluster population. This gradient indicates that the thick disk clusters formed out of material which was enriched by previous generations of star formation, whereas the halo clusters show no general trend of metallicity with radius, indicating the absence of a connection between dissipation and enrichment for at least some of the halo population (see Chapter 3).

Outside the Local Group, the first suggestions of color gradients were made by the Stroms and their collaborators in their study of the GCSs of Virgo ellipticals (e.g. Strom *et al.* 1981). These were based on photographic plates which offered the large areal coverage critical for studies of radial gradients. However, plates are not ideal for photometry of faint sources, so this evidence for gradients was insecure. The first studies of elliptical galaxy GCSs using CCDs were also inconclusive on the issue of color gradients. These were hampered by the small size of the early generation CCDs.

Studies utilizing large format CCDs have now established that most or all elliptical galaxy GCSs have color gradients such that they are redder in the center and bluer at large radii. Some of the published examples are NGC 1399 (Bridges *et al.* 1991; Ostrov *et al.* 1993), M87 (Lee and Geisler 1993), NGC 3923 (Zepf *et al.* 1995a), NGC 3311 (Secker *et al.* 1995), and NGC 5813 (Hopp *et al.* 1995). The best data at the time of writing are those of Geisler *et al.* (1996) on the NGC 4472 GCS. Each of these studies covers a similar radial range, which is roughly 5–25 kpc. The data for NGC 4472 are shown in Figure 5.10. In terms of metallicity, the gradients for these galaxies are all

$$\Delta[\text{Fe/H}]/\Delta \log r \simeq -0.5 \text{ dex}, \qquad (5.6)$$

with typical uncertainties of about 0.1–0.2 dex. Although there are also a number of examples in the literature in which no gradients are found, these are generally consistent with the typical gradient given above. This is because these studies have not had the long color baseline and good radial coverage that are required to obtain significant results for GCS color gradients at this level.

It is interesting to compare the GCS color gradients with those of the underlying galaxy light. Typical color gradients for the integrated light of ellipticals correspond to metallicity gradients of

$$\Delta[Fe/H]/\Delta \log r \simeq -0.2 \text{ dex} \tag{5.7}$$

(e.g. Davies *et al.* 1993 and references therein). The integrated light gradients appear to be somewhat smaller for high-luminosity ellipticals, which includes most of those for which GCS color gradients have been measured. Thus, it seems that the GCS color gradients are steeper than those in the integrated light. This comparison is not completely straightforward because gradients in integrated light are typically determined from the center out to about an effective radius, whereas the GCS gradients are not measured in the center and reach to much larger radii. However, a steeper color gradient for the GCS relative to the integrated light is observed in the cases with a significant overlap in the observed radial range of the two gradients, as in the example shown in Figure 5.10.

Theoretical implications of these results are discussed in the following chapter. One observational implication is that the color difference between the GCS and the integrated light introduced in the previous section will depend on the radius at which it is measured. Although the difference is sufficiently large that it is apparent at all radii covered by current observations, the size of the effect is dependent on whether both colors are taken at the same radius and, if they are the same, on the choice of radius at which the comparison is made.

5.7   Color distributions

As the quality of the data on the colors of a large number of objects in extragalactic GCSs has improved, it has become possible to extract more information about these colors than simply their mean or median. This additional information leads to correspondingly better constraints on models for the formation and evolution of GCSs and their host galaxies. As an example, for the Galactic system, the metallicity (color) distinction between the thick disk population and the halo population is a critical element in understanding Galactic evolution. It is now possible to obtain colors of a large number of globular clusters around galaxies at distances of tens of Mpc with sufficient accuracy to carry out similar analyses for the GCSs of these galaxies.

5.7.1   *Dispersion of distributions*

The dispersions of the color distributions of GCSs provide information on the age and metallicity spread between globular clusters. In the early 1990s, improvements in the photometric precision of observations of the GCSs of giant ellipticals in Virgo and Fornax revealed that their GCS color distributions were broader than those of the Milky Way and M31 (Couture *et al.* 1991; Geisler and Forte 1990). The breadth of these distributions was often characterized by a Gaussian dispersion in

metallicity, which is typically about $\sigma[\text{Fe/H}] = 0.6$ dex for the giant ellipticals and about $\sigma[\text{Fe/H}] = 0.4$ dex for the Local Group spirals. These dispersions are about twice the photometric errors. Although the distributions are non-Gaussian, more sophisticated techniques also indicate a larger scale for the color distributions of giant ellipticals compared to Local Group spirals.

A second important point is that the color distributions are broad at all radii. This result can clearly be seen in the plot of color against radius for the NGC 4472 GCS shown in Figure 5.10, and was first noted by Couture *et al.* (1991). Thus, a large color dispersion is intrinsic to elliptical galaxy GCSs and is not caused by a color gradient in a sample covering a range in radius. In fact, the large intrinsic dispersion of the color distribution is a primary reason it is difficult to detect color gradients in GCSs.

5.7.2 Detectability of peaks in color distributions

Through deeper multicolor photometry of GCSs, it has become possible to go beyond a determination of the mean color and dispersion and study the *shape* of color distributions. The shape of these distributions reflects the enrichment history of the GCS and, by extension, the galaxy itself. As usual, one is limited by the problem that broad-band colors reflect both age and metallicity variations. Fortunately, models in which elliptical galaxies are enriched during a single collapse and a burst of star formation predict that the metallicity distribution should be single-peaked (see Chapter 6). Moreover, the population is essentially coeval, so that this metallicity distribution will be directly reflected in the observable color distribution. In contrast, other models in which star formation in ellipticals is episodic tend to predict peaks in the color distribution due to metallicity and age differences between stars produced in different bursts.

The basic observational requirements for obtaining useful shape information on the color distribution include a broad wavelength baseline for age/metallicity sensitivity and good photometric precision. Without these factors, the observed color distribution will tend to be dominated by photometric errors and will appear Gaussian, irrespective of the intrinsic shape of the distribution. As an illustrative case, we consider the requirements for distinguishing a color distribution which is a single Gaussian from one which is a composite of two Gaussian populations. (As we discuss in Chapter 6, this situation is also theoretically motivated.) The required color precision and sample size are set primarily by the expected difference between the mean colors of the two populations and the intrinsic width of the individual color distributions. A simple expectation is that two peaks can be detected only if they are separated by at least twice the standard deviation of the individual peaks (Everitt and Hand 1981). There are many factors which mitigate this optimstic estimate of the required photometric precision (Ashman *et al.* 1994). For example, Ashman *et al.* (1994) showed that, with sample sizes of 100–200, which are typical of currently available data, a photometric precision of a factor of 2.5–3.0 less than the expected separation is required for a reliable detection or rejection of bimodality using the best available algorithms. For large sample sizes of roughly 500–1000 objects, the required precision is about a factor of two less than the expected separation.

To place these numbers in the context of current observations, we adopt the case of two roughly coeval populations with no instrinsic metallicity dispersion and a separation of $\Delta[Fe/H] = 1.0$. For typical conversions of $\Delta[Fe/H] = 1.0$ to color indices (e.g. Couture *et al.* 1990; Worthey 1994), the required photometric precision to detect or reject bimodality reliably with a sample size of 100–200 objects is $\sigma(V - I) \leq 0.07$, or $\sigma(B - I) \leq 0.13$. As the photometric precision improves, one limit for the detection of individual populations is the internal metallicity dispersion of the populations themselves. In the case of the halo globular clusters of the Galaxy, the observed dispersion is about $\sigma[Fe/H]= 0.33$ (Armandroff and Zinn 1988), or $\sigma(B - I) = 0.12$.

These numbers are only illustrative, since it is not obvious *a priori* what the metallicity and age differences between the populations will be. The case described above is loosely based on the predictions of Ashman and Zepf (1992) for a merger picture for elliptical galaxy formation, and is similar to the difference between the halo and thick disk clusters in the Galaxy. We note that, in many cases, the effect of including both age and metallicity differences will be to make the two populations harder to distinguish, because the younger population will often be the more metal-rich population as it will have had more time to undergo enrichment. However, in general, stellar populations are much more sensitive to metallicity differences than age differences once they are older than one or two Gyr (e.g. Worthey 1994; Charlot and Bruzual 1996).

It is also not obvious *a priori* that there will be only one or two populations instead of three or many more. At some point a large number of small populations with widely varying ages and/or metallicities will resemble a single large population. Thus the question that can be answered is whether there are one or two, or perhaps several, dominant populations with significant color differences. These populations can themselves be composites of populations with similar ages and metallicites. Regardless, the detection of distinct populations of clusters in the color distribution clearly points to distinct episodes of cluster formation and, by implication, galaxy formation as well.

5.7.3　First results on peaks in color distributions

The first statistical test for bimodality in the color distributions of elliptical galaxy GCSs was presented by Zepf and Ashman (1993). They utilized the best available photometry at the time, which was 72 $(B - I)$ colors for NGC 4472 globular clusters (Couture *et al.* 1991), and 60 $(C - T_1)$ colors for NGC 5128 globular clusters (Harris *et al.* 1992). Histograms of these data are shown in Figure 5.11. Based on a mixture-modeling analysis, Zepf and Ashman (1993) found that a bimodal distribution was a better fit than a unimodal one at a 98.5% confidence level for the NGC 4472 GCS and a 95% confidence level for the NGC 5128 GCS.

Neither the histograms shown in Figure 5.11, nor the reported confidence limits, provide conclusive evidence for bimodality in these datasets. The limiting factor is the small number of objects in each sample. Fortunately, the advent of large-format CCDs has allowed bigger samples to be obtained. The first studies to exploit these larger CCDs include those of Geisler and collaborators, who analyzed the color distribution of the GCSs of NGC 1399 (Ostrov *et al.* 1993), and M87 (Lee and Geisler 1993). In both cases, a statistical analysis indicated that the color distributions were bimodal or multimodal, with significantly better fits for two or more Gaussians than

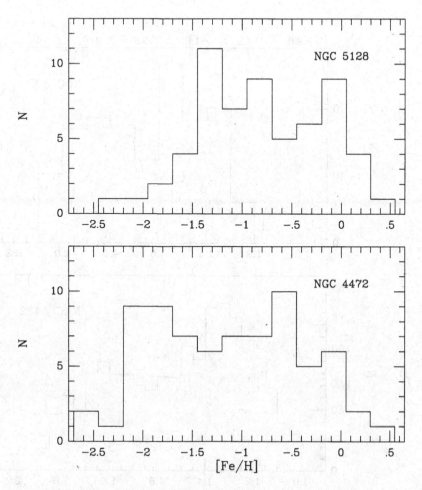

Fig. 5.11. Color distributions for the GCSs of NGC 4472 and NGC 5128 from Zepf and Ashman (1993). These provided the first evidence for bimodality, since confirmed by better data.

a single Gaussian distribution. Similarly, Ajhar *et al.* (1994) surveyed a number of early-type galaxies, and obtained $(V - I)$ colors for 100 or more clusters around five Virgo ellipticals. A statistical analysis of these data showed that two of the GCSs are bimodal with high confidence, one is likely to be bimodal, and two appear to be unimodal (Zepf *et al.* 1995a).

Further evidence that bimodal color distributions are common for the GCSs of elliptical galaxies came from the studies of NGC 3311 by Secker *et al.* (1995) and of NGC 3923 by Zepf *et al.* (1995a). Figure 5.12 shows histograms of the color distributions of the NGC 3923 GCS and the NGC 4472 GCS from Geisler *et al.* (1996). In addition to color distributions which are not well fit by a single Gaussian, it is notable that the NGC 3923 GCS appears to be shifted redward relative to the NGC 4472 GCS. As these observations were taken in the same photometric system, ·

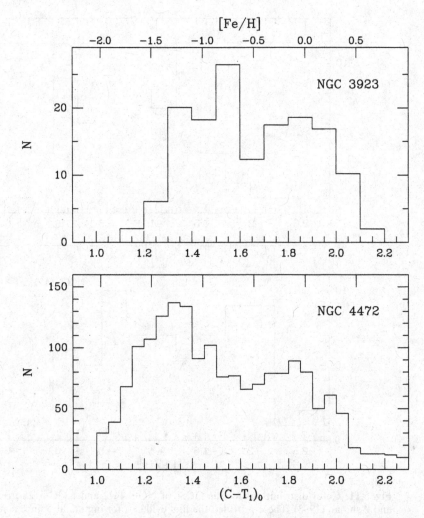

Fig. 5.12. Color distributions for the GCSs of NGC 3923 and NGC 4472 from Zepf *et al.* (1995a) and Geisler *et al.* (1996) respectively. NGC 4472 is more luminous than NGC 3923, but its GCS is bluer and presumably more metal-poor, indicating that there is significant scatter about any relationship between galaxy luminosity and GCS color.

and the calibrations were checked against the colors of the galaxies themselves, the effect appears to be real. This result suggests that, at least in one case, the scatter in mean color is the result more of a shift in the color distribution as a whole rather than a difference in the relative numbers of clusters in two peaks with fixed mean colors. It also suggests that the location of the 'blue' peak in elliptical galaxy GCSs varies by up to several tenths of a dex in terms of [Fe/H]. A final point is that Zepf *et al.* (1995a) showed that the detection of two or more peaks in the color distribution is not significantly affected by non-linearities in the color–metallicity relation predicted by theory.

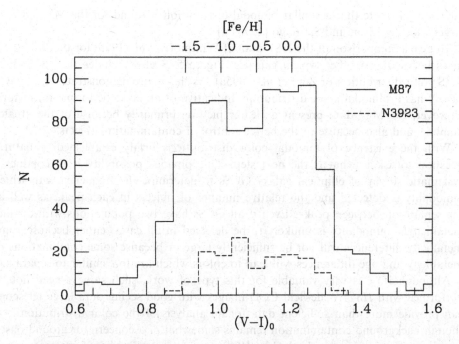

Fig. 5.13. Color distributions for the GCSs of M87 from Whitmore *et al.* (1995) and NGC 3923 from Zepf *et al.* (1995a). The color distributions of both GCSs are bimodal, although the number of clusters observed in M87 is much greater.

5.7.4   *Present status*

The above results suggest that the color distributions of at least some elliptical galaxy GCSs are bimodal, which has important implications for galaxy formation models, as discussed in Chapter 6. However, there are several ways in which the data can be improved. In particular, significantly increasing the number of clusters with good photometry allows a more detailed study of the individual peaks. Deeper data also probe intrinsically fainter clusters to insure that they follow trends established for the brighter end of the population. Moreover, there is some background contamination in ground-based data, which, even if it is at a low level, is still a source of noise.

Dramatic improvements in all these areas were achieved in the analysis of *WFPC2* images of the M87 GCS by Whitmore *et al.* (1995). They showed with unprecedented clarity that the color distribution of the M87 GCS divides into two distinct populations, as shown in Figure 5.13. One reason for the clear visual indication of two distinct peaks is the large number of clusters detected (~ 1000) with high photometric precision. This wealth of clusters is due in large part to the depth which can be reached with *HST* imaging of compact sources. An additonal factor is the richness of the M87 GCS itself, which makes for a high surface density of objects. The *WFPC2* images also marginally resolve globular clusters at the distance of Virgo, providing a clean distinction between the clusters and background galaxies, which are much more extended, and even foreground stars, which are point sources. It is also interesting to note that a similar bimodal distribution is found for the M87 GCS at larger radii by Elson and Santiago (1996a,b).

To give a comparison of the *HST* data of Whitmore *et al.* (1995) for the M87 GCS to earlier work on other typical galaxies, Figure 5.13 also includes the NGC 3923 GCS color distribution of Zepf *et al.* (1995a). As this figure demonstrates, each GCS clearly has a bimodal color distribution, indicative of an episodic formation history. However, the M87 data present a clearer picture, primarily because of the greater numbers and also because of the better control of contaminating objects.

With the existence of bimodal color distributions firmly established, a natural question to ask is what is the next step. One obvious possibility is to conduct a systematic survey of elliptical galaxy GCSs to determine the frequency with which bimodality is detected and the relative number of objects in each peak, as well as the separation between peaks. Even if all GCSs have two populations with distinct metallicities, bimodality is unlikely to be detected in all cases, either because some metallicity differences will not be sufficiently large or because some combinations of metallicity and age differences will lead to colors which are too similar to separate.

Although *HST* is very valuable for this type of work, such surveys need not be performed with *HST*. Wide-field CCD images with good seeing on a large telescope can provide more than sufficient data for an analysis of the color distribution. Although background contamination remains somewhat of a concern for ground-based imaging surveys, it can be checked by *HST* imaging or ground-based spectroscopy of a selected subset of the whole survey. One example of the capability of ground-based studies is the Geisler *et al.* (1996) analysis of the NGC 4472 GCS. In roughly three hours of imaging through the Washington C and T_1 filters with a 4 m telescope, they detected well over 1000 clusters with typical precision in terms of [Fe/H] of 0.15 dex. With these data, Geisler *et al.* (1996) clearly confirm the previous detection of bimodality in the NGC 4472 GCS by Zepf and Ashman (1993).

Geisler *et al.* (1996) were also able to begin to constrain the radial profiles of the 'red' and 'blue' populations separately. They showed that at least some of the color gradient of the system as a whole is due to the changing ratio of red to blue populations, such that the blue population makes up a higher fraction of the total at large radii than at small radii. Similar constraints for the GCSs of other galaxies are likely to become available soon. Although any single *HST WFPC2* frame does not cover a large radial area, there are several pointings for some galaxies, which can be combined for a more sensitive study of radial variations.

Deep *HST* images that reveal GCSs with bimodal color distributions can also be used to compare the GCLFs of the two populations. As described in Section 5.3, for systems with identical *mass* functions, those which are more metal-poor will be brighter in blue bandpasses (Ashman *et al.* 1995). Whitmore *et al.* (1995) made an initial attempt at such a comparison, fitting a Gaussian to the 'red' and 'blue' populations individually. They found that the blue population was brighter in V by 0.13 ± 0.19 magnitudes. They also found that the distance modulus determined from the I-band data was 0.05 magnitudes larger than that found in V. Both these are consistent with model expectations given a constant mass function, but do not conclusively demonstrate that this is the case. Methods which directly compare the observed GCLFs of the two populations without Gaussian fitting may give more conclusive results, and, of course, more data will be very useful.

5.8 Spectroscopy

Spectroscopy of globular clusters around galaxies outside the Local Group was near or beyond the limit of the capabilities of large telescopes until the mid-1990s. Despite the technical difficulties, these studies attracted interest because of the unique and valuable information they provide. Now, with the availibilty of efficient wide-field, multi-slit spectrographs and larger telescopes, it is feasible to obtain spectroscopy of large numbers of objects in the GCSs of galaxies outside the Local Group. We expect that these advances will have a major impact on the subjects we describe below.

Spectroscopy offers several important advantages over broad-band colors. Perhaps the most basic advantage of spectroscopy is that redshifts from spectra establish a candidate cluster's association with its parent galaxy on an individual basis. In photometric studies, this association is always statistical in nature, although the probability that a given object is a globular cluster of a specific galaxy can be very high. One of the primary motivations for spectroscopy is that radial velocities determined from the spectra provide a valuable probe of the mass distribution of the galaxy. The extended spatial distribution of GCSs makes them particularly useful in this regard, since they probe the outer regions of early-type galaxies where it is difficult to obtain other kinematical information.

Another motivation for spectroscopy is that the observed strengths of stellar absorption lines provide important information about the stellar populations of the globular clusters. Spectra of modest resolution and signal-to-noise can be used to estimate metallicities, given an adopted age and calibration based either on models or on observations of Galactic clusters. These estimates, which are similar to those derived from broad-band colors, are very useful because they are derived independently and are insensitive to some concerns which affect colors, such as reddening. One example of the utility of samples of line strengths is that they can be used as an independent test of the existence of distinct populations in the GCSs observed in the color distribution. Moreover, color and/or line strength information can be combined with the observed velocities to examine the kinematics of the different populations in a GCS. With higher-quality spectra of globular clusters, one can also study other lines and line ratios to place further constraints on the stellar population of the GCS. For example, the [Mg/Fe] ratio is of particular interest because it is observed to be greater than solar in the integrated spectra of ellipticals (Worthey *et al.* 1992; Davies *et al.* 1993). This result places interesting constraints on formation models because Mg is produced solely by type II supernovae, whereas Fe is produced in type I supernovae as well. Thus [Mg/Fe] ratios reflect the stellar initial mass function and the timescale of star formation.

Although the programs completed to date are modest compared to the goals described above, a number of interesting results, particularly relating to evidence for dark halos around ellipticals galaxies, have emerged. Mould *et al.* (1990) studied the globular cluster systems of M87 and NGC 4472. Based on velocities for 43 globular clusters around M87 extending out to a radius of 8′, they measured a velocity dispersion of $\sigma = 386 \pm 42$ km/s, with a rotation of $v_{rot} = 108 \pm 54$ km/s, where σ is *not* corrected for rotation. For this galaxy, they combined the GCS kinematics at large radii with the stellar kinematics at small radii to conclude that the data are inconsistent with a constant mass-to-light (M/L) model, and require a dark halo around M87. Mould *et al.* (1990) also obtained velocities for 26 clusters around NGC 4472 out to a radius of 5', and found a velocity dispersion of $\sigma = 340\pm50$ km/s, with a rotation of $v_{rot} = 113\pm57$ km/s. These data also suggest the presence of a dark halo around NGC 4472, but at a lesser confidence level than for M87. Their estimates for the median metallicity of the GCSs for both galaxies, based on the Mg lines and a Galactic cluster calibration, are consistent with the photometric estimates of about [Fe/H] = -0.9 (Bridges and Hanes 1992). The M87 GCS was also studied by Brodie and Huchra (1991), who used a number of stellar absorption lines to estimate the metallicity and found a result consistent with that given above.

The GCS of NGC 1399, the giant elliptical at the center of the Fornax cluster, was the subject of a spectroscopic study by Grillmair *et al.* (1994). They obtained velocities for 47 globular clusters around the galaxy out to a radius of 9', and found a dispersion of $\sigma = 388 \pm 54$ km/s, with no evidence of rotation. This velocity dispersion is at least as high as the central stellar velocity dispersion, and requires that M/L increase with radius.

A spectroscopic study of the GCS of the more nearby elliptical galaxy NGC 5128 has also been undertaken. Velocities of 87 globular clusters are given in Harris *et al.* (1988). Based on these velocities, Hui *et al.* (1995) presented evidence that the blue (metal-poor) population identified by Zepf and Ashman (1993) does *not* rotate, while the red (metal-rich) population has significant rotation like that seen in the planetary nebulae and stellar kinematics. This is the first example outside the Local Group of kinematic confirmation of a population distinction based on globular cluster colors.

More recently, Bridges *et al.* (1997) studied the GCS of NGC 4594, an Sa galaxy known as the Sombrero because of its very bright bulge and a striking dust lane. They determined velocities for 34 clusters out to a radius of 5.5' (14 kpc for a distance to NGC 4594 of 8.55 Mpc), and found a velocity dispersion of 265 ± 29 km/s, with a possible rotation of $v_{rot} = 73\pm45$ km/s, where the given σ is uncorrected for rotation. Bridges *et al.* (1997) used the projected mass estimator (Bahcall and Tremaine 1981) to determine that $(M/L)_V = 16$ within 14 kpc with the 90% confidence interval ranging from $(M/L)_V = 11$ to 21.5. Adopting the extreme case of circular orbits reduces this by about a factor of 1.5. Even in this case, the M/L estimate from the GCS at large radius indicates a rising M/L ratio when combined with the results of Kormendy and Westphal (1989), who found $(M/L)_V \simeq 4$ at small radii, with evidence for a gentle rise to about $(M/L)_V \simeq 8$ at 3', where all M/L values have been normalized to the same distance. Bridges *et al.* (1996) also used the G-band spectral feature at 4300 Å and the calibration of Brodie and Huchra (1990) based on Galactic globular clusters, to estimate a mean metallicity for the NGC 4594 of [Fe/H] = -0.7 ± 0.3, consistent with earlier photometric estimates (Bridges and Hanes 1992).

5.9   Systems of young globular clusters

The previous sections of this chapter are concerned with the properties of systems of globular clusters roughly similar to those in our Galaxy. In particular, these globular clusters are generally old like those in our Galaxy, or at least they are not obviously young in that their colors and luminosities are inconsistent with very young ages of less than 1 Gyr or so. However, not all objects with masses and radii like those of Galactic globular clusters are old. The most nearby examples are the handful of massive, 'populous young star clusters' in the LMC, discussed in Chapter 4. The focus of this section is on the search for, and discovery of, significant numbers of candidate young globular clusters in galaxies outside the Local Group, particularly in merging galaxies, which are known to have vigorous starbursts, and which were predicted to form young globular clusters (e.g. Schweizer 1987; Ashman and Zepf 1992).

As described in Chapter 4, predicting what globular clusters are expected to look like when they are young is straightforward. They should be bright, blue, and compact (e.g. Ashman and Zepf 1992). The compactness is a basic characteristic of globular clusters, without which they would not survive within the tidal field of their host galaxy for a Hubble time. The bright luminosity and blue color comes from the fact that young stellar poplations have massive stars, which are bright and blue. The luminosity and color evolution can be predicted using models of stellar populations. As an example, Figure 5.14 shows the evolution with time of the absolute B magnitude and $(B - V)$ color for an instantaneous burst of star formation of mass of 2×10^5 M$_\odot$, assuming a Miller–Scalo (1979) stellar initial mass function and the models of Bruzual and Charlot (1995), which are for solar metallicity. This figure shows that young globular clusters are several orders of magnitude brighter and substantially bluer than old stellar populations such as the Galactic globular clusters. Precise predictions of color and luminosity depend on the IMF and metallicity, but generally high luminosities, blue colors, and compact sizes are the identifying signatures of candidate young globular clusters.

Although it is straightforward to predict what objects such as the Galactic globular clusters looked like in their youth, it is difficult to determine if any individual young star cluster will survive and evolve over another Hubble time to be known to future observers as a globular cluster. This problem is discussed in depth in Chapter 4. It is clear that the presence of compact, young star clusters is a necessary condition for the recent formation of globular clusters, even though it is not a sufficient one. The observational task is to determine if objects consistent with the properties of young globular clusters are found in various galaxies, and then to determine if these objects are likely to evolve to become globular clusters like those in the Milky Way or M31.

With the exception of the few massive, populous young clusters in the LMC and several other nearby dwarf galaxies discussed in Chapter 4, the first system of bright, blue candidate young globular clusters identified were those in the galaxy merger NGC 3597 by Lutz (1991). However, this ground-based study was only able to place upper limits on the sizes of the objects which were much greater than the few pc typical of Galactic globular clusters. The identification of these objects as candidate young globular clusters was clearly tentative.

The study of young globular cluster systems underwent a dramatic revolution with *HST* imaging, which allows much tighter limits on the sizes of candidate objects. *HST* imaging is also very efficient at detecting sources with expected sizes on the order of 0.1″. The first discovery of a large number of objects with the properties expected of young globular clusters was made by Holtzman *et al.* (1992), who found about 50

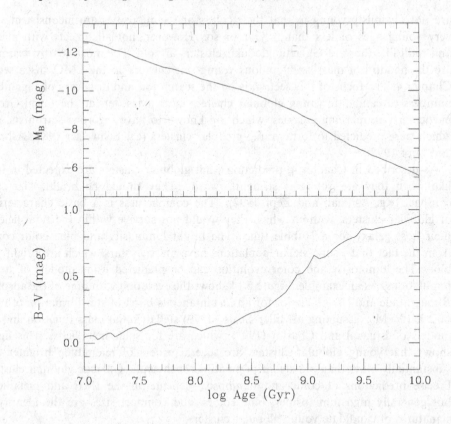

Fig. 5.14. This figure shows the evolution of a single-burst stellar population with a mass of 2×10^5 $M_\odot$. The top panel is a plot of absolute B magnitude against time in Gyr, and the lower panel is a plot of $(B - V)$ color against age. The general trend of brighter magnitudes and bluer colors is common to all stellar population models. This figure is specifically based on Bruzual and Charlot (1995) models with a Miller–Scalo (1979) stellar initial mass function.

objects in the peculiar galaxy NGC 1275 with $\langle M_V \rangle \simeq -13$, $\langle (V-R) \rangle \simeq 0.3$, and sizes of less than about 10 pc. The blue colors and bright magnitudes were subsequently confirmed with imaging at the Canada–France–Hawaii Telescope by Richer *et al.* (1993), who found an average $(B-V)_0$ color of about 0.1.

Following this initial discovery, Whitmore *et al.* (1993) obtained *HST* images of the prototypical merger NGC 7252 and found a population of bright, blue, compact objects very similar to those in the peculiar galaxy NGC 1275. Because NGC 7252 is a well-known galaxy merger, the discovery of candidate globular clusters in this galaxy provided strong support for the prediction that globular clusters form in galaxy mergers. The connection between galaxy mergers and candidate young globular clusters has now been confirmed by subsequent studies of merging systems utilizing the refurbished *HST*. These include observations of the NGC 4038/4039, or the 'Antennae' (Whitmore *et al.* 1995), NGC 3597 (Holtzman *et al.* 1996), NGC 3921 (Schweizer *et al.* 1996), and NGC 3256 (Zepf *et al.* 1996). Figure 5.15 is an *HST* image

Fig. 5.15. An image of the galaxy merger NGC 3256 taken with the *Hubble Space Telescope*. This image reveals a number of bright, compact objects. A quantitative analysis of exposures in *B* and *I* filters indicates that these have the colors and sizes expected of young globular clusters.

of the last galaxy. There are no known examples of a galaxy with star formation associated with a merger which does not have candidate young globular clusters.

These discoveries have opened the door for a closer study of the properties of these candidate young clusters, aimed at determining whether these objects are truly globular clusters. One useful tool is spectroscopy, which provides much better contraints on the stellar population of a cluster, and gives direct evidence that an object is a star cluster associated with its host galaxy. With these goals in mind, moderate resolution spectroscopy of the two brightest candidate young globular clusters in NGC 7252 (Schweizer and Seitzer 1993) and the brightest cluster in NGC 1275 (Zepf *et al.* 1995b) was obtained. In all three cases, the spectra support the identification of the objects as young globular clusters. Firstly, redshifts from the spectra clearly place all the clusters at the distance of their host galaxy. Morever, the spectra all show very strong Balmer absorption lines, indicative of a stellar population younger than 1 Gyr and older than the age of massive stars hot enough to produce an HII region. A comparison of the spectra with stellar population models and young stars and star clusters of known ages shows that the most likely ages range from about 0.1–0.9 Gyr for the cluster in NGC 1275 (Zepf *et al.* 1995b) to 0.01–0.5 Gyr for the clusters in NGC 7252 (Schweizer and Seitzer 1993). These age estimates are consistent with those from broad-band colors. The spectra also show metallic absorption lines, albeit at the weak level expected in young stellar populations. Zepf *et al.* (1995b) attempted to use these lines to constrain the metallicity of the young globular cluster, and found that is *very* roughly solar. Similar conclusions were reached by Fritze-v. Alvensleben and Burkert (1995) in an anlaysis of the spectrum of the brightest cluster in NGC 7252.

A second approach is to obtain deep imaging of the systems of candidate young clusters to provide a complete census of the objects down to faint magnitudes. This is useful both for studying the luminosity function of the young cluster system and for determining the total number and luminosity of compact young clusters formed. The original images of NGC 1275 and NGC 7252 were limited to fairly bright objects because of the flawed optics of the pre-refurbishment *HST* and because of the distance of these galaxies. By studying the nearby merging system, NGC 4038/4039, Whitmore *et al.* (1995b) were able to observe more intrinsically faint clusters and discovered that these objects are very numerous. In total, they found over 700 clusters, with absolute magnitudes from $M_V = -15.5$ to -9.5, and the number distribution following a power law in luminosity of $N(L) \propto L^{-1.8}$ over the whole observed range.

An even greater number of compact young star clusters was observed in NGC 3256 by Zepf *et al.* (1997), using images from the refurbished *HST*. Like the NGC 4038/4039 systems (and the LMC), the number distribution is roughly a power law in luminosity with a slope of -1.8, although Zepf *et al.* (1997) find some evidence for a flattening of the slope at about $M_B \sim -11$. The colors show a wide spread in $(B-I)$, caused at least partially by reddening from dust, but are generally very blue and suggest ages on the order of 10^8 years.

The power-law luminosity functions found for the young cluster systems in these nearby galaxy mergers are clearly different than the log-normal GCLFs found in galaxies like the Milky Way and M87. Instead, they are fairly similar to the luminosity function of the LMC clusters. This shape of the luminosity function raises interesting questions about the relationship of these compact young clusters to old GCSs (e.g. van den Bergh 1995c). However, the youth of these mergers and their globular cluster systems make the comparison difficult to interpret. Meurer (1995) points out that when the ages of the clusters are similar to the age of the starburst as a whole, it is possible to obtain a power-law luminosity function from a log-normal mass function.

The physical reason for this effect is that luminosity evolution is most rapid for the youngest ages, leading to a small number of very bright clusters and a much larger number of somewhat fainter clusters. Even given an initial mass function which is a power law (as expected in some models of globular cluster formation discussed in the following chapters), various dissolution and destruction mechanisms will act to change this mass function over time. Moreover, most of these mechanisms act to remove low-mass clusters preferentially, thus leading to possible evolution from a power law towards something resembling a log-normal GCLF (Murali and Weinberg 1996; also Section 3.5).

HST imaging programs have also revealed that populations of compact, bright, blue objects are not restricted to galaxy mergers alone, but appear in a number of galaxies which have strong starbursts and no evidence of galaxy interactions or mergers. These include several nearby starburst galaxies (e.g. O'Connell *et al.* 1994, 1995; Meurer *et al.* 1995; Watson *et al.* 1996), and galaxies with nuclear starburst rings (e.g. Barth *et al.* 1995). These data suggest that the common physical condition for the production of the dense, young star clusters is a strong starburst.

Another conclusion to arise from these studies is that a substantial fraction of the star formation in the observed bursts appears to be occurring in the compact young clusters. The most striking example is NGC 3256, for which the ~ 1000 compact blue objects make up 19% of the B and 7% of the I light within the central $40'' \times 40''$ region (roughly $r \leq 3.5$ kpc). The NGC 4038/4039 system has about half this fraction of light in its young cluster system (Zepf *et al.* 1997). The numbers can be even higher than the global NGC 3256 value within smaller starburst regions in these galaxies and others. Therefore, even with a significant loss of clusters through dissolution and destruction, these data suggest that the formation efficiency of candidate young globular clusters is high. The implications of this high efficiency of cluster formation are discussed in the following two chapters.

The spatial distribution of the candidate young globular clusters can also be delineated by these imaging studies. In all systems studied to date, their distribution appears to follow that of the underlying galaxy, or specifically the starburst region in those cases in which the starburst is confined to the nucleus or a ring. In dynamically evolved mergers, such as NGC 7252 and NGC 3921, the candidate young globular clusters have radial profiles like that of the galaxy light (Whitmore *et al.* 1993; Schweizer *et al.* 1996). A similar pattern seems to hold for younger mergers such as NGC 3256 and NGC 4038/4039 (Zepf *et al.* 1997; Whitmore *et al.* 1995), where the clusters appear to roughly follow the blue light distribution. In the nuclear starburst cases, the very young objects observed appear to be located within the previously identified starburst regions (e.g. Barth *et al.* 1995; Meurer *et al.* 1995).

The properties of the systems of candidate young globular clusters discovered in starbursts in merger and other environments are very much like those expected of globular cluster systems at young ages. This agreement does not imply that every object will evolve to become a globular cluster, and in fact the luminosity function and efficiency arguments suggest that many of them will dissolve or be destroyed. However, some of the candidate young globular clusters, particularly the most massive, are likely to evolve over the next Hubble time into objects like the globular clusters of the Milky Way and M87. Determining whether the evolution of the young cluster systems results in a mass function like that seen in old cluster systems remains an outstanding question. One way to begin to address this question is to compare the young globular cluster systems at different ages, and presumably different stages of evolution.

6

Globular clusters and galaxy formation

The study of globular cluster systems (GCSs) has long been motivated, at least in part, by the idea that these systems can be used as fossil records of the formation history of their host galaxies (e.g. Harris and Racine 1979; Harris 1991). As described in the previous chapter, empirical information concerning GCSs has grown tremendously in both quantity and quality in recent years. This growth has led to more discriminating tests of models of the formation and evolution of galaxies through the properties of their globular cluster systems. Understanding galaxy formation and evolution is one of the primary challenges in extragalactic astronomy and cosmology. In this chapter, we describe models of galaxy formation and the constraints placed on these models by observations of globular cluster systems.

6.1 Models of galaxy formation

6.1.1 Galaxy properties

In the search for a physical model of galaxy formation and evolution, one of the primary questions is why galaxies have such a wide variety of morphologies, star formation histories, and stellar kinematics. One specific issue of great interest is why some galaxies are ellipticals which have old stellar populations, are dynamically hot, and follow de Vaucouleurs' surface brightness profiles (see Section 5.2), and others spirals which have been forming stars at roughly a constant rate over a Hubble time in a rotationally supported, exponential disk.

The dynamical differences between ellipticals and spirals are interesting, particularly since these galaxies have roughly similar mass densities. This similarity suggests that spirals and ellipticals collapsed by comparable amounts from their protogalactic stage. Theories of structure formation predict that protogalaxies acquire angular momentum through tidal torques caused by neighboring protogalaxies. The initial specific angular momentum has some dispersion, but is expected to be roughly the same for all protogalaxies with a value similar to that of current-day giant ellipticals (e.g. Peebles 1969; Efstathiou and Jones 1979; Heavens 1988). For spirals, a collapse of a factor of ten can account for both their rotation and their relative overdensities (Fall and Efstathiou 1980). The critical question is how elliptical galaxies avoided spinning-up since they appear to have collapsed by the same amount.

A second key question is why the mean age of the stellar population of a galaxy correlates so well with its dynamics. Whatever mechanism produces galaxies with low specific angular momentum must be tied to star formation history in order that elliptical galaxies lack the ongoing star formation characteristic of spiral galaxies. Elliptical galaxies also follow tight correlations between their dynamical properties and stellar populations, providing further evidence for a close connection between galaxy formation and star formation and enrichment. Specifically, there is little scatter about the trend for stronger metallic absorption lines and redder colors (indicative of higher metallicities and/or older ages) for ellipticals which are more luminous or have higher-velocity dispersions (indicative of larger mass). These relationships have been explored in many papers and reviews, beginning with Faber (1973) and Sandage and Visvanathan (1978). For elliptical galaxies, similarly tight constraints can be placed on stellar population variations by considering the 'Fundamental Plane' of elliptical galaxies (e.g. Dressler *et al.* 1987; Djorgovski and Davis 1987; Faber *et al.* 1987, and many subsequent papers).

6.1.2   *Galaxy formation in isolation*

One place to start in the consideration of galaxy formation is with an individual galaxy, forming and evolving in isolation. In addition to simplicity, this scenario is also natural *if* galaxies extend only as far as their observed optical extents and are weakly clustered. In this case, the frequency of galaxy collisions is expected to be low, as given roughly by:

$$N_{merg} \sim 2 \times 10^{-4} \left(\frac{n}{0.02h^3 \text{ Mpc}^{-3}} \right) \left(\frac{r}{10 \text{ kpc}} \right)^2 \left(\frac{v}{300 \text{ km/s}} \right) \left(\frac{t}{10^{10} \text{ yrs}} \right), \quad (6.1)$$

where n is the density of galaxies, r the typical radius of a galaxy, and v the typical relative pairwise velocity of galaxies (see also Tremaine 1981). As we will show later, this rate appears to underestimate the frequency of galaxy collisions significantly, because of both the extended dark halos of galaxies and their intrinsic clustering.

The landmark work in the study of the formation of an individual galaxy is Eggen, Lynden-Bell and Sandage (1962; hereafter ELS), who focussed on the Milky Way. As discussed in Chapter 3, they argued that the halo of the Galaxy formed out of a gaseous protogalaxy in which star formation proceeded rapidly, on the order of the free-fall time. The gas left over from this process then dissipated into a disk and formed stars at a gradual rate.

This general picture of dissipational collapse appears to provide a description of the formation of spiral galaxies which accounts for many of their properties. As described earlier, the overdensities and angular momenta of spiral galaxies can be understood if they have collapsed by about a factor of ten. More specific details about stellar populations and chemical enrichment can be understood with additional considerations of gas infall and enrichment from the halo population.

The ELS scenario for the formation of the Galactic halo can be extended to provide a description of the formation of spheroidal systems in general, thereby including elliptical galaxies. Ellipticals are the galaxies in which rapid star formation in the spheroidal component continued until all the gas was turned into stars, leaving no gas out of which to form a slowly evolving disk. Lynden-Bell (1967) showed that 'violent relaxation' of a stellar system naturally produces the $r^{1/4}$ law that fits elliptical galaxy surface brightness profiles. Such a process is likely to occur if star formation is rapid and the galaxy consequently collapses on a free-fall timescale. In the 1970s, several proposals for why ellipticals formed stars much more rapidly than spirals were put forward, including differences in initial angular momentum (Sandage, Freeman, and Stokes 1970), and initial density differences (Gott and Thuan 1976). However, both these hypotheses have been subsequently shown to be inconsistent with broader cosmological considerations (e.g. Silk and Wyse 1993).

The inclusion of hydrodynamical processes in a model of the evolution of an individual galaxy was pioneered by Larson. In a series of papers (Larson 1969a,b, 1974a,b, 1975), he utilized a heuristic treatment of star formation, the subsequent return of metals and energy into the interstellar medium, and chemical enrichment, to make model predictions for the properties of galaxies, particularly elliptical galaxies. This work has subsequently been expanded upon by a number of authors, including Carlberg (1984a,b), Arimoto and Yoshii (1987), and Matteucci and Tornambè (1987). All these models assume that the gas collapses dissipationally and forms stars which then enrich the gas out of which more stars form. This continues until the energy input from supernovae is strong enough to drive a galactic wind, which terminates the formation process.

Comparison of these models to observations indicate that they are able to account for the typical metallicity of elliptical galaxies and the trend that more massive ellipticals are more metal-rich. More specifically, Arimoto and Yoshii (1987) and Matteucci and Tornambè (1987) found that for some choices of the dark matter distribution and the stellar IMF they were able to reproduce the relationship between color and luminosity observed for elliptical galaxies. However, in doing so, they predicted that the [Mg/Fe] ratio observed in the stars of ellipticals would be less than solar and would decrease with increasing galaxy mass. This prediction is in direct contradiction with observations which showed that [Mg/Fe] ratios are greater than solar for ellipticals, and are larger for more massive ellipticals (Worthey *et al.* 1992; Davies *et al.* 1993). These observations require that massive ellipticals are enriched much more by type II supernovae relative to type I supernovae than expected from the models described above, as might occur if these galaxies had a stellar initial mass function biased towards high mass stars (e.g. Matteucci 1994; Elbaz *et al.* 1995; Zepf and Silk 1996). Color gradients also naturally result from dissipational models, which can be taken as a success since they are often observed in elliptical galaxies (e.g. Kormendy and Djorgovski 1989). However, quantitatively, the predicted gradients (e.g. Carlberg 1984b) are larger than those observed. These models also produce larger gradients for more massive ellipticals, contrary to observation (e.g. Kormendy and Djorgovski 1989).

The most persistent weakness of these models is that none of them provide a natural explanation for the observation that bright ellipticals usually have little or no rotation (see reviews by Binney 1981 and Illingworth 1981), whereas faint ellipticals have more significant rotation (Fall 1979; Davies *et al.* 1983, and subsequent papers). As discussed in the previous section, galaxy formation models must account for the lower specific angular momentum of ellipticals relative to spirals while at the same time explaining the similar mass densities. In models of galaxy formation in isolation, these differences between spirals and ellipticals must be established at the protogalactic epoch. As mentioned above, simple approaches such as the assumption that ellipticals are protogalaxies with initially low angular momentum and/or higher densities either fail to reproduce the observed properties of ellipticals or are inconsistent with cosmological constraints (Silk and Wyse 1993 and references therein). More promising models are based on efficient angular momentum transfer in protoellipticals which dumps angular momentum in the outer regions of these galaxies where there are few observational constraints. One specific proposal which has the merit of making a link to other properties of elliptical galaxies is that protoellipticals are lumpier than protospirals (Silk and Wyse 1993). It is claimed that this lumpiness leads to efficient angular momentum transfer, and to more rapid star formation.

Some elements of the dissipation and enrichment in the single collapse picture for the formation of elliptical galaxies must appear in any model in order to account for color gradients and the relationship between mean metallicity with galaxy mass. However, there are several reasons for looking beyond single collapse models. As emphasized above, the low angular momentum of massive ellipticals remains an outstanding question not directly answered by these models. Moreover, when extended dark halos are taken into account, the estimated frequency of galaxy collisions is significantly greater than the estimate given in equation (6.1). A number of observational and theoretical results point to an important role for galaxy mergers in the formation and evolution of galaxies. In the following section, we explore models in which the formation and evolution of galaxies are significantly affected by the interactions and mergers between them.

6.1.3   *Hierarchical structure formation and galaxy mergers*

One long-standing hypothesis for the origin of galaxy morphology is that all galaxies begin as disk galaxies, and elliptical galaxies are formed by subsequent mergers of these disk systems (Toomre 1977; Toomre and Toomre 1972). One of the primary arguments in favor of this model is that the process of galaxy transformation by mergers appears to be observed locally. Specifically, Toomre (1977) identified 11 nearby systems as the mergers of two disk galaxies at various dynamical stages in the merging process. Subsequent observational studies of these systems (e.g. Hibbard and van Gorkom 1996), combined with numerical simulations of the merger process (e.g. Barnes and Hernquist 1992), have confirmed that galaxy mergers provide a good explanation for the properties of these systems.

Detailed studies of the more dynamically evolved Toomre systems indicate that these objects have surface brightness profiles characteristic of elliptical galaxies (e.g. Schweizer 1982). These observations appear to confirm the expectation that mergers of disk galaxies can lead to spheroidal systems as a result of dynamical relaxation occurring during the merger (Toomre and Toomre 1972). The violent relaxation associated with the merger is also expected to erase much of the dynamical 'memory' that the stars have of their circular orbits in the progenitor spiral disks. Moreover, most of the angular momentum from the orbits of the two merging galaxies is lost to their extended halos. Therefore, the merger of two rotationally supported disk galaxies can lead to a slowly rotating elliptical galaxy.

The colors and spectra of the oldest two of the Toomre objects (NGC 7252 and NGC 3921) indicate that these galaxies are post-starburst objects (Schweizer 1982; Schweizer *et al.* 1996). This is consistent with the idea that galaxy mergers induce bursts of star formation, which consume much of the available cold gas and heat up the rest. Therefore, it is plausible that the resultant merger remnant will have little ongoing star formation, and will passively evolve into an old stellar population.

In this picture, spiral galaxies are those systems which have not experienced strong interactions or mergers with other galaxies of similar mass, at least not since the current disk formed. This naturally accounts for the thiness of their disks and their ongoing star formation. The thiness of spiral disks can be used to place constraints on their merger history (e.g. Toth and Ostriker 1992), although these constraints depend on physical parameters such as the density of the infalling galaxies (e.g. Walker *et al.* 1996; Quinn *et al.* 1993).

This view of galaxy evolution is generally consistent with popular cosmological models such as the cold dark matter (CDM) scenario in which structure forms through hierarchical clustering (Peebles 1984; Blumenthal *et al.* 1984). In such models, small scales turn-around from the Hubble Flow first, so that galaxies are built from the 'bottom up'. As the universe evolves, small-scale structures are subsumed within larger protogalaxies. While this hierarchical clustering is somewhat distinct from genuine *merging*, galaxy mergers are a key feature of such models. For example, merging is a critical element in galaxy evolution for models which adopt CDM-dominated universes and prescriptions for the gas cooling, star formation, feedback, and dynamical interactions between galaxies (e.g. Cole *et al.* 1994; Kauffmann *et al.* 1994; also White and Rees 1978).

6.2   Globular clusters in monolithic collapse models

As described in Chapter 5, elliptical galaxy GCSs are more spatially extended and have bluer colors than the integrated light of their host ellipticals. In the Galaxy, the majority of globular clusters are associated with the halo, an old and spatially extended component of the Galaxy (see Chapter 3). These properties of GCSs were some of the first to be observationally established, and suggested a natural link between globular cluster formation and the early stages of galaxy formation. Initially, the observations were interpreted primarily in the context of models in which galaxy formation and evolution occurs in isolation. The bluer colors of globular clusters relative to the underlying galaxy light were regarded as being due to lower metallicities. Combined with the flatter profiles of GCSs, this result was used to suggest that globular clusters formed somewhat earlier than the bulk of the stars in these galaxies (de Young *et al.* 1983; Harris 1986; Hanes and Harris 1986). The color gradients in GCSs around ellipticals (Section 5.6) are also expected in models involving dissipational collapse and enrichment.

One possible alternative explanation for the extended spatial distribution of GCSs is that globular clusters close to galaxy centers are preferentially destroyed through dynamical processes (see Sections 3.5 and 7.1; also Murali and Weinberg 1996). Although these dynamical processes must be at work at some level, there are several observations which indicate that they are not responsible for the blue, extended nature of elliptical galaxy GCSs. Firstly, it is difficult for the destruction mechanisms to be effective out to radii of tens of kpc, at which differences in the slope of the radial profiles are observed (see Section 3.5, and Aguilar *et al.* 1988). Secondly, dynamical destruction mechanisms tend to lead to a variation of the GCLF with radius, which has not been observed in the limited number of systems that have been studied (Sections 3.5 and 5.3.2). Finally, it is difficult for destruction mechanisms to reproduce the bluer color of the GCS compared to the integrated light of the elliptical galaxy at the same radius. Thus, the blue, extended nature of elliptical galaxy GCSs is likely to reflect conditions at the time of their formation rather than subsequent dynamical evolution.

Although these observations are consistent with a dissipational collapse model, they do not provide a compelling argument for such a model. Color gradients point to some role for dissipation, but do not require that it occur in a single collapse. There is also no clear physical model for the preferential formation of globular clusters *before* the bulk of the stars in dissipational collapse pictures, as is required to explain the differences in surface brightness profiles and metallicities of globular clusters and field stars. Some of the globular cluster formation models described in Chapter 7 include the idea that globular clusters form only when the metallicity of the protogalactic gas is low. In such scenarios, it is natural to expect that the majority of globular clusters form before the bulk of field stars. Once enrichment raises the metallicity, globular cluster formation is inhibited in such pictures. Unfortunately, as we discuss in Chapter 7, such models appear to be inconsistent with observation. Thus an earlier formation epoch for globular clusters relative to stars must currently be assumed, without a firm physical basis, in dissipational collapse models.

Observations also show that brighter elliptical galaxies have GCSs which are more metal-rich than those of fainter spiral galaxies (see Figure 5.7 and Section 5.5). A trend for higher-luminosity galaxies to have more metal-rich GCSs can be accounted for in monolithic collapse models as a result of the deeper potential wells of more luminous galaxies retaining more metals (e.g. van den Bergh 1975; Brodie 1993; Forbes *et al.* 1996). However, Figure 5.7 shows that there may be a discontinuous jump in GCS metallicity from spirals to ellipticals (e.g. Perelmuter 1995; Ashman and Bird 1993). Moreover, Figure 5.8 demonstrates that any trend of GCS metallicity with elliptical galaxy luminosity appears to have significant scatter (e.g. Zepf *et al.* 1995). As discussed below, these observations may be interpreted as favoring a merger model.

The relative specific frequency of globular clusters (S_N) around ellipticals and spirals has played an influential role in using the properties of GCSs to constrain models of galaxy formation (e.g. van den Bergh 1990a). As discussed in Section 5.4, ellipticals have roughly 2–3 times as many globular clusters per unit stellar mass as spirals (i.e. their T values are 2–3 times larger; Zepf and Ashman 1993). Although the dependence of S_N (or T) on L means that this ratio will depend somewhat on the luminosities of the galaxies being compared, it seems likely that the greater number of globular clusters around ellipticals relative to spirals is a real effect. The argument against forming ellipticals by mergers is thus straightforward: ellipticals have too many globular clusters to be formed by combining spirals (van den Bergh 1990a). This reasoning implicitly assumes that the number of globular clusters remains constant. Clearly, a merger model can account for the observations if globular clusters form in mergers. While there was originally little evidence for the formation of GCs in mergers, observations of ongoing galaxy mergers now reveal signifcant populations of candidate young globular clusters (Section 5.9). Moreover, the hypothesis that a component of elliptical galaxy GCSs was formed in mergers is testable, as is described in the next section.

6.3   A merger alternative

Driven by various lines of evidence which suggest that mergers are an important part of elliptical galaxy formation, Ashman and Zepf (1992) considered the expected properties of the globular cluster systems of elliptical and spiral galaxies in the context of a simple merger model in which ellipticals are produced by the merger of two spirals. The most basic ingredient of this picture is that, in addition to the globular clusters of the progenitor spirals, a population of globular clusters forms in the merger itself. The observational evidence for this formation is discussed in Section 5.9, and models of globular cluster formation in starbursts and mergers are detailed in the following chapter.

Because T values are typically 2–3 times as high in ellipticals as spirals, Ashman and Zepf (1992) proposed that roughly an equal number of globular clusters were likely to be formed in the merger as were present in the progenitor spirals. They showed that the formation of this number of globular clusters in mergers was reasonable in terms of the available gas mass in the progenitor galaxies and plausible globular cluster formation efficiencies. A second important element is that the disks of the progenitor spirals are expected to have undergone dissipation and enrichment since the formation of their halo GCSs. Thus the material out of which globular clusters and other stars form during the merger has a smaller spatial extent and a higher metallicity than the halos of the progenitor spirals.

As a result, the globular cluster systems of elliptical galaxies formed in this simple merger model have two populations of globular clusters (Ashman and Zepf 1992). The halo clusters of the progenitor spirals form one population which is metal-poor and spatially extended. A second population is composed of globular clusters formed during the merger. These clusters are metal-rich and spatially concentrated. The more spatially extended distribution of the metal-poor halo clusters naturally gives rise to an overall metallicity gradient in elliptical galaxy GCSs. Moreover, this halo population 'puffs up' the overall GCSs radial profile relative to the stars, most of which form after some dissipation in the progenitor disks or during the merger. Similarly, the low metallicity of the halo population causes the GCS to have a lower metallicity on average than the integrated light of the elliptical galaxy at a given radius.

The formation of a metal-rich population of cluster during the merger which makes the elliptical leads to a higher mean metallicity for elliptical galaxy GCSs relative to spiral galaxy GCSs (Ashman and Zepf 1992). Thus the merger model naturally accounts for the observation that elliptical galaxy GCSs have higher metallicities than those of spiral galaxies (Section 5.5). Although it is difficult to separate the effects of morphology from luminosity in the current sample, Perelmuter (1995) finds that the data are better fit by the simple merger model than the simple collapse picture (see also Ashman and Bird 1993).

The ability of the merger model to reproduce the basic observational characteristics of GCSs around ellipticals stems from the phase of enrichment and dissipation between the formation of globular clusters around the progenitor spirals and those formed in the merger. This enrichment and dissipation seems fairly inevitable, but it is *not* sufficient in itself to account for the observed properties of GCSs around ellipticals. An additional requirement is that the ratio of stars to globular clusters that form out of the enriched gas must be higher than the same ratio in the halos of the progenitor spirals. This is necessary to avoid overproduction of globular clusters in the resulting ellipticals (the ratio of globular clusters to stars in the Milky Way *halo* corresponds to a specific frequency of around 25). Further, if stars and globular clusters form in the same ratios from both the halo and enriched gas, the GCS of the resulting elliptical will have the *same* profile as the underlying galaxy.

In the context of the merger model, it is natural to suppose that metal-rich globular clusters and stars form in the merger, whereas the progenitor disks form only stars. This is empirically reasonable, since the disks of present-day spirals do not seem to produce globular clusters, whereas violent mergers produce a large number of globular cluster candidates efficiently. Thus even if the ratio of stars to globular clusters formed in the merger is similar to that in the halos of spirals, the lack of globular cluster formation in the progenitor disks ensures that the merger model can reproduce the properties of elliptical galaxy GCSs outlined above.

While the preferential formation of stars in the progenitor disks is a necessary part of the merger picture, it raises the question of whether there is sufficient gas left over in the disks to make the required number of globular clusters in the subsequent merger. One approach is to consider the efficiency of globular cluster formation, where efficiency is defined as the mass fraction of *available* gas that forms globular clusters (see Ashman and Zepf 1992). Unfortunately, this line of attack is hampered by our ignorance of the evolution of protogalactic gas in spirals. For example, if all the material now in the Milky Way disk and halo was originally in halo gas, this total mass might be considered the available gas mass. In this case, the formation efficiency of halo globular clusters is less than about 10^{-4}. Alternatively, if only the mass now in the Milky Way halo is considered, the formation efficiency of globular clusters is around 10^{-2}, or even higher if some of the field halo stars formed in globular clusters. With this range of possible efficiencies, it is perhaps not surprising that the merger model can account for the 'additional' globular clusters required by observations of specific frequencies of elliptical galaxies (Ashman and Zepf 1992).

Fortunately, there are other considerations and observations that have the potential to eliminate some of the current uncertainties in the merger model. One example is the suprasolar alpha-element ratios of stars in elliptical galaxies which indicate that ellipticals cannot be formed by simply merging two stellar disks like that of the Milky Way. Thus the progenitor spirals are required to be fairly gas-rich at the time of the merger in order to provide a way to form stars with suprasolar abundances (Section 6.1.1). The fact that most elliptical galaxies have old stellar populations is consistent with this requirement, since it implies that the vast majority of mergers occurred at relatively high redshifts when the gas content of spirals is likely to be higher than the present day. Finally, if it is demonstrated that the young globular cluster candidates in galaxy mergers are indeed young globulars, it will be possible to determine the efficiency of globular cluster formation in mergers directly. This would be an important step in refining the merger model.

We are left with a scenario in which ellipticals form from the mergers of fairly gas-rich spirals. These spirals have their own populations of metal-poor halo globular clusters and have likely undergone some star formation in their young disks. The formation efficiency of globular clusters is poorly constrained, but the available gas mass is sufficient to produce the required number of globular clusters in the merger to explain the specific frequency of ellipticals.

6.4   Distinguishing between models

The previous sections of this chapter have shown that either a single collapse or a merger model can explain the radial profiles, colors, and color gradients observed in GCSs. The obvious goal is to find observations which distinguish between these two models. As described above, elliptical galaxies formed by mergers will have GCSs composed of two or more distinct populations of globular clusters (Ashman and Zepf 1992). The halo clusters of the progenitor spirals form one population which is metal-poor and spatially extended. A second population is composed of globular clusters formed during the merger. These clusters will be younger, and also more metal-rich than the halo population, since they form out of material which has been enriched in the disks of the progenitor spirals and in the merger itself. Because most mergers are likely to occur at moderate or high redshifts, metallicity differences play a much larger role than age differences in determining the colors and metallic absorption-line strengths for typical elliptical galaxy GCSs (Zepf and Ashman 1993). Thus, in most cases the clusters associated with the original spirals are expected to be bluer and have weaker metal lines than the clusters formed during the merger in which the elliptical galaxy formed.

This prediction of two or more observably distinct populations in the GCS of an elliptical galaxy formed by a merger is in direct contrast to the single population predicted by monolithic collapse models. A cluster population formed in a monolithic collapse is expected to be roughly coeval and have a smooth, single-peaked metallicity distribution. For example, the metallicity distribution of stars in elliptical galaxies arising from a simple closed-box model is plotted in Figure 6.1 (cf. Binney and Tremaine 1987). The shape of this distribution closely resembles that found in more detailed models (e.g. Arimoto and Yoshii 1987).

Observations of the distribution of colors or line strengths for GCSs of elliptical galaxies thus provide a way to distinguish between monolithic collapse and merger models of the formation of elliptical galaxies. Although the formation of galaxies is probably much more complex than the scenarios described above, the general trends seem secure. More than one peak in the color distribution indicates episodic formation as expected in a merger picture, and single-peaked distributions are suggestive of monolithic collapse models.

As discussed in Section 5.7, bimodal color distributions have been discovered and confirmed in the GCSs of several giant elliptical galaxies. The observations are therefore consistent with the predictions of merger models, and appear to rule out a single, monolithic collapse as the formation mechanism for their host ellipticals. Although the number of confirmed bimodal cases is still not large, the high rate of discovery in well-studied systems suggests that they are not rare.

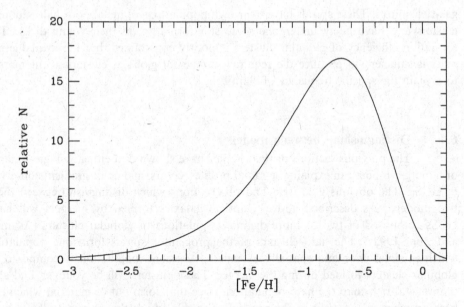

Fig. 6.1. The metallicity distribution of stars produced by a simple closed-box model.

6.5   Problems and possibilities

While bimodal color distributions provide considerable evidence in favor of a merger origin for elliptical galaxies, the merging picture, at least in its simplest form, has difficulty in explaining a few of the detailed observations of the GCSs of ellipticals. The purpose of this section is to describe some of these problems and to discuss how a more sophisticated merger picture might compare to the observations. We also discuss possible modifications to single-collapse models which might enable such pictures to reproduce bimodal color distributions. We emphasize that these ideas are speculative, but the quality of current data means that we are in a position to go beyond basic tests of the two competing scenarios.

6.5.1   *Observational conflicts with the merger model*

The simple merger picture leads to a number of testable predictions (Ashman and Zepf 1992), most of which have been confirmed, as described above. However, one prediction which is in conflict with observation is that the mean metallicity of the GCSs of ellipticals is expected to increase with increasing specific frequency S_N (or T). The reason for this prediction is that, in the merger picture, the higher S_N values of ellipticals relative to spirals result from the metal-rich globular clusters formed in mergers. Since S_N increases with the proportion of metal-rich clusters one expects a correlation between S_N and mean metallicity. This prediction requires the additional assumptions that the GCSs of ellipticals are comprised of just two populations, one from the progenitor spirals and one produced in the merger, and that the mean metallicities of these two populations are the same for all ellipticals. Figure 5.9 shows a plot of S_N against mean metallicity for elliptical galaxies where both quantities have been determined. It is apparent that there is no obvious trend of the kind expected in the simple merger scenario.

A comparison between the GCSs of specific elliptical galaxies highlights this and related problems with the simple merger model and points to where the simplifying assumptions break down. The GCSs of M87, NGC 3311, NGC 3923 and NGC 4472 have all been imaged in the Washington photometric system (see Section 5.5), so a comparison of their mean colors and color distributions is particularly informative. M87 and NGC 4472 have similar inferred mean metallicities, but M87 has a specific frequency roughly a factor of three higher than NGC 4472. A comparison of the color distributions of these galaxies reveals that the blue and red peaks occur at metallicities of around $[Fe/H] \simeq -1.5$ and -0.5 in both cases, and that the ratio of metal-poor to metal-rich clusters is around unity for both galaxies. In the context of the simple merger picture, one would expect M87 to have a much higher fraction of metal-rich clusters than NGC 4472, reflecting its higher specific frequency.

The situation becomes more confusing when one considers the GCSs of NGC 3311 and NGC 3923. The former galaxy is rather similar to M87 in terms of its high specific frequency and location in a cluster. NGC 3923 has a specific frequency more typical of giant ellipticals and similar to that of NGC 4472. Despite these similarities, mean metallicities of the GCSs of NGC 3311 and NGC 3923 are higher than those of NGC 4472 and M87. The higher mean metallicity is *not* driven by a larger proportion of metal-rich clusters. Instead, the metallicity distributions of the GCSs of NGC 3923 and NGC 3311 have the appearance of being similar to that of NGC 4472 but shifted to higher metallicity by about 0.4 dex. The remarkable aspect of these results is that the four galaxies under consideration neatly divide into two pairs on nearly all properties except the metallicity distributions of their GCSs. M87 and NGC 3311 are classic high-S_N galaxies, whereas NGC 3923 and NGC 4472 are bright ellipticals with normal values of S_N. However, in terms of GCS metallicity, NGC 3923 and NGC 3311 have similar distributions, as do M87 and NGC 4472. This peculiar state of affairs is difficult to explain in either the merger or single-collapse model.

Another issue is that given the redward shift of the NGC 3311 and NGC 3923 GCSs, the *blue* peak of these GCSs is clearly *redder* than the peak of the halo clusters of the Milky Way and M31. This suggests that NGC 3311 and NGC 3923 do not possess metal-poor clusters originally associated with spiral galaxies. However, both NGC 3311 and NGC 3923 have bimodal metallicity distributions indicative of a merger origin.

The M87 system illustrates another problem with the simple merger picture. The comparable numbers of metal-rich and metal-poor clusters in this galaxy imply that the specific frequency contributed by the metal-poor clusters is $S_N \simeq 7$. This is higher than the specific frequency of spiral galaxies (even accounting for differences in stellar populations), but in the merger picture, these metal-poor clusters are supposed to have been originally associated with spirals.

6.5.2   Can the merger model be saved?

The central question raised by these observations is whether they rule out the merger model altogether, or whether they point to failures in the simplifying assumptions of the model. In either case, the observations hold important clues to the nature of a complete model of galaxy formation that can also account for the properties of globular cluster systems.

The most basic oversimplification of the Ashman and Zepf (1992) picture is that elliptical galaxies are clearly not the merger of just two spirals. Many ellipticals, particularly those whose GCSs have been studied, are simply too massive. More generally, cosmological models and computer simulations indicate that many giant galaxies undergo a complex merger history (e.g. Kauffmann 1996). To a first approximation, this complication does not alter the basic prediction of the merger model that the GCSs of ellipticals are comprised of two or more distinct populations. At some stage the model requires spirals to merge to produce an elliptical, but if these ellipticals go on to merge with other galaxies, one still anticipates metal-poor clusters from the spirals and metal-rich clusters formed in the gas-rich mergers. One likely outcome is that in some cases the multiple populations will wipe out evidence for bi- or multimodality in the metallicity distributions, so that even in the merger picture there will be *some* GCSs where multiple populations are not expected to be detected.

While the basic predictions of the simple merger model seem robust, a more complex merger history may help to explain the detailed predictions. For example, cD galaxies such as M87 are expected to have undergone a large number of mergers and to have accreted many dwarf galaxies. As discussed in Chapters 4 and 5, dwarf galaxies typically have metal-poor globular clusters and S_N values greater than those of spirals. It is therefore plausible that a significant fraction of the metal-poor clusters currently surrounding M87 were orginally associated with dwarfs rather than spiral galaxies, in which case the high specific frequency of this population is less problematic.

Such accretion events also have a bearing on the predicted S_N–[Fe/H] relation of the simple merger picture which is not found observationally. If the fraction of globular clusters associated with accreted dwarfs varies between giant ellipticals, scatter will be introduced into the predicted S_N–[Fe/H] relation. However, a potentially more important effect is that, contrary to the assumption of the simple picture, the metal-poor and metal-rich peaks in the GCSs of ellipticals do *not* have universal values. A scatter in the mean metallicities of the metal-poor and metal-rich populations will tend to wipe out the increase of the mean [Fe/H] of the combined system with S_N, although it seems unlikely that the effect could completely eliminate the trend.

Another issue is that there is considerable scatter in the S_N values for spiral galaxies (see Section 5.4), even for spirals of the same morphological type. This can also dilute the predicted S_N–[Fe/H] relation of the simple merger picture. As an example, consider two spirals with high values of S_N which merge, producing a typical number of metal-rich globulars in the merger. The S_N of the resulting globular cluster system will be somewhat higher than average, but the mean metallicity will be somewhat lower. In other words, the expectation of an increase of [Fe/H] with S_N can be reversed under certain conditions. While such events are, by definition, not the norm, they will clearly occur and tend to wipe out the trend predicted by the simple merger picture.

The lack of metal-poor globular clusters in galaxies like NGC 3311 and NGC 3923 is an extreme manifestation of the variation in peak metallicity of the metal-poor globular clusters in ellipticals. The primary problem for the merger scenario is the lack of clusters with metallicities typical of those found for the halo globulars of spiral galaxies. If galaxies like NGC 3923 and NGC 3311 experienced, at some stage in their past, a major merger involving spiral galaxies, where are the metal-poor clusters associated with those spirals? This question is sharpened by the fact that both GCSs exhibit metallicity bimodality which is generally regarded as the most compelling evidence in favor of the merger picture.

To address this question, we note that the assumption of a universal metallicity for metal-poor clusters was prompted by its simplicity, as well as some observational evidence. The halo globular clusters of the Milky Way and M31, as well as the old clusters of the LMC, have mean metallicities around $[Fe/H] \simeq -1.6$. Since the number of spirals with well-determined GCS metallicities is small, one way out of the conundrum presented by NGC 3311 and NGC 3923 is to suppose that some spiral galaxy GCSs have significantly higher metallicities. We emphasize that there is currently little if any evidence for this suggestion (the bulge-dominated spiral M104 has a metal-rich GCS, but may be a better candidate for the result of a merger rather than a typical merger progenitor), but it may be necessary to reconcile the merger model with the absence of metal-poor globular clusters around some ellipticals. It is therefore of considerable interest to obtain more metallicity distributions of spiral galaxy GCSs.

One of the difficulties in carrying out detailed tests of the merger scenario is that we have little information concerning how globular cluster systems have evolved since their formation (see Sections 3.5 and 7.1). Of particular relevance is the possibility that dynamical processes have destroyed a large number of globular clusters, so that the systems observed today are comprised of the survivors of an initially much larger population. Such complications were not considered by Ashman and Zepf (1992), but their inclusion has the potential to modify the predictions of the merger scenario quantitatively. As an example, Murali and Weinberg (1996) have carried out numerical simulations of the effects of various destructive processes on globular cluster systems. These authors reach the intuitively reasonable conclusion that in galaxies with lower (central) stellar densities globular clusters survive more easily. The brightness and stellar density of giant ellipticals are anti-correlated, so these results suggest that destruction of globular clusters is more effective in faint ellipticals and spirals than in bright ellipticals. Not only do these results provide an explanation for some of the increase in S_N with elliptical galaxy luminosity, they suggest that the number of clusters in spirals, which are generally metal-poor, may have been significantly depleted (see also Section 3.5). However, when such spirals merge and their globular clusters end up in a giant elliptical galaxy, destruction processes are likely to be less effective, so that higher specific frequencies for this population may result. As mentioned in other chapters, the observed invariance of the GCLF with radius in galaxies such as the Milky Way and M87 is difficult to reconcile with the destruction of a large fraction of globular clusters.

The purpose of the above discussion is to emphasize that the merger model, in the form outlined by Ashman and Zepf (1992), does not account for certain characteristics of the GCSs of elliptical galaxies. Whether refinements to the basic model, such as including the effects of destructive processes and multiple mergers, can resolve the current discrepancies remains to be seen. The lack of constraints on parameter space means that it is quite easy to imagine such refinements, but without observational motivation such an excercise is of limited value. However, the increasing amount of information on GCS color distributions, gradients, kinematics, and other properties suggests that such modeling will be an active area of research in the next few years.

6.5.3   *Another look at monolithic collapse models*

In the light of the detailed observations discussed above, it is clearly important to assess how well rival pictures to the merger scenario reproduce the observed properties of GCSs. While there is currently no detailed alternative model, it is possible to investigate the broad scenario in which giant ellipticals and their GCSs form from a single, probably lumpy, protogalactic cloud. Given current cosmological results, a complete model will likely include some degree of merging. Thus the operational distinction between such a model and the merger scenario described above is that in monolithic collapse models galaxy mergers do *not* produce a significant population of globular clusters in the GCSs of giant ellipticals.

Color bimodality in elliptical galaxy GCSs is the strongest evidence in favor of the merger scenario, particularly since it was a *prediction* based on the merger model. A central question is therefore whether a monolithic collapse can reproduce this bimodality. It is extremely difficult to see how this could occur if there is only a single phase of such a collapse, as has generally been assumed in the past. Generically, it appears that globular cluster color bimodality demands two phases of globular cluster formation. Thus, in the context of monolithic collapse models, one requires a mechanism to 'turn off' globular cluster formation before chemical enrichment has proceeded far, with a subsequent mechanism to 'turn on' globular cluster formation again once gas in the collapsing galaxy has been sufficiently enriched. One possibility is that, after an initial phase of globular cluster formation, galactic winds suppress star formation and simultaneously boost the metal content of the surviving galactic gas (e.g. Berman and Suchkov 1991). This possibility has not been investigated specifically in terms of the GCS it might produce.

Durrell *et al.* (1996a) discuss this topic in the context of the Harris and Pudritz (1994) model of globular cluster formation, in which the progenitors of globular clusters are 'Super Giant Molecular Clouds' (SGMCs). Durrell *et al.* (1996a) suggest that there is an initial burst of globular cluster formation in giant ellipticals responsible for the metal-poor clusters that uses up a relatively small fraction of the total available gas. The remaining enriched gas is then assumed to undergo dissipative collapse and form further SGMCs which produce the metal-rich globular clusters. While this idea can in principle explain the color bimodality and different spatial characteristics of the metal-poor and metal-rich globular clusters, Durrell *et al.* (1996a) offer no physical mechanism for the hiatus in globular cluster formation. A mechanism is also required to prevent the enriched gas from spinning-up as it dissipates.

An important observational constraint on monolithic collapse scenarios is provided by the mean metallicity of GCSs around ellipticals (Section 5.5). The increase in the mean metallicity with galaxy luminosity from dwarf to giant ellipticals has traditionally been cited as evidence in favor of monolithic collapse (see van den Bergh 1975; Brodie 1993, and references therein; Forbes *et al.* 1996). The idea is that more luminous galaxies have deeper potential wells and are therefore better able to retain their metals. Consequently, collapsing gas in these deeper wells is enriched to higher levels. However, the *scatter* in the mean metallicity of GCSs around ellipticals (Section 5.5) presents a problem for this picture. At a given luminosity, observations indicate that GCSs of ellipticals have been enriched to a range of metallicities. There is no strong correlation *for giant ellipticals* between GCS metallicity and galaxy luminosity.

This state of affairs is understandable in the merger model: the mean metallicity of a GCS will depend on the detailed merging history of the host galaxy and no simple correspondence between mean metallicity and galaxy luminosity is expected. In monolithic collapse pictures, this observation is harder to understand. At the very least, it suggests that the enrichment history of a GCS is not solely determined by the depth of the potential well of the host galaxy. It is conceivable that the scatter in the mean metallicity can be explained in monolithic collapse models if globular clusters form in discrete overdense lumps within the protogalaxy, but such a model has not yet been developed.

The observed color bimodality requires that any successful model of elliptical galaxy formation must somehow incorporate two (or more) epochs of globular cluster formation. Bimodality or multimodality is an inevitable consequence of the merger model, although, as discussed in the previous section, there are some detailed observations which are inconsistent with the simple variant of the merger picture. In monolithic collapse models there are as yet unexplored avenues that may be able to produce episodic globular cluster formation, as well as the other properties of GCSs that are currently difficult to understand in such scenarios. We discuss future observations that may help to resolve these issues in the final chapter.

7

The formation of globular clusters

At present, there is no widely accepted theory of globular cluster formation. In this chapter, we describe some of the general ideas that have been proposed in this area, and compare these ideas with the constraints placed on globular cluster formation models by the observations described in earlier chapters.

One piece of evidence that has played an important role in the development of this field is the lack of current globular cluster formation in the Milky Way today. Open clusters and stellar associations form quite happily at the present epoch in the Galactic disk, but globular clusters do not. Until relatively recently, a significant fraction of astronomers would have probably argued that globular cluster formation was something that *only* occurred in the early universe. Others have claimed for some time that the most massive young star clusters found in the Large Magellanic Cloud and other similar environments are genuine analogs of the old globular clusters of the Milky Way and other galaxies. As discussed in Chapter 5, there is now evidence that globular clusters are currently forming in merging and interacting galaxies. While it is premature to regard this evidence as conclusive, there seems to be a growing acceptance of the idea that globular clusters can, under certain circumstances, form at the present epoch. As we show below, this possibility has significant consequences for models of globular cluster formation.

7.1   Globular cluster destruction

At the risk of putting the cart before the horse, we first discuss the possibility that old globular cluster systems are the survivors of an initially much more significant population. One motivation for this approach is that it is important to know to what degree globular cluster systems have evolved since the time of formation. This is essential in order to use the observed properties of such systems to constrain formation scenarios. A second motivation is the attempt to determine whether disrupted globular clusters can account for a significant fraction of field stars in the Galactic halo. We considered the dynamical evidence for this hypothesis in Chapter 3. While studies indicate that such disruption *may* be efficient enough to disrupt a sufficient number of globular clusters, the uncertainties in the numerical modeling leave the question open. The purpose of this section is to complement the results gleaned from dynamical considerations with observational comparisons between halo field stars and globular clusters. We also extend the discussion to galaxies other than the Milky Way.

An additional reason for considering globular cluster destruction at this stage is that the process is required by many models of globular cluster formation. As discussed below, certain models predict that the initial mass spectrum of globular clusters should continue down to masses well below those observed for the old globular clusters of the Milky Way and other galaxies. Thus a central issue is whether it is possible to remove low-mass clusters from the initial distribution over a Hubble time. The most effective long-term dynamical process for low-mass cluster destruction appears to be evaporation, accelerated by tidal shocks (Section 3.5). Such clusters are also vulnerable to disruption at relatively young ages through stellar mass loss and supernovae (discussed further below). In the specific case of evaporation, it is worth noting that the observed decrease of globular cluster density with galactocentric radius in the Milky Way and other galaxies suggests that clusters closer to the galaxy center are more susceptible to disruption. (An increase in globular cluster radius at a fixed mass leads to an increase in the evaporation timescale – see equation (3.6).) This is expected to lead to a decrease in the lower mass limit of globular clusters with increasing galactocentric radius, assuming that the initial globular cluster population extended to low masses at all galactocentric distances. In the case of the Milky Way, there is a hint of such a trend (see Figure 3.8), but the data are far from conclusive.

In Chapter 3 we noted that the spatial distribution of stars and globular clusters in the Milky Way halo are similar, but that the metallicity distributions and elemental abundance patterns in the two populations show significant differences. These comparisons appear, at first sight, to pose a serious problem for the idea that the current population of field stars represents disrupted globular clusters. Basically, if the hypothetical initial population of globular clusters were comprised of objects with properties similar to the current population, the disruption of most of this initial population would *not* produce the field star population we observe today.

There are, however, ways to avoid such a conclusion. In the case of CN abundances, the discussion of Chapter 2 indicates that CN-rich stars may be the product of stellar evolutionary processes that are only effective within globular clusters. If most of the hypothetical population of globular clusters are disrupted fairly early in their lifetime, these stellar evolutionary processes may not have had sufficient time to produce CN-rich stars. The difference between the metallicity distributions of stars and globular clusters presents more of a problem. If all the current halo field stars initially formed in globular clusters, it appears that the initial globular cluster population had somewhat different metallicity characteristics to the surviving population. This requires some mechanism to preferentially disrupt the lowest metallicity globular clusters. Models in which the observed globular clusters of the Milky Way are the survivors of an initially larger system must therefore address this question.

Combining the constraints from dynamical studies with observations of metallicity distributions and elemental abundances seems to suggest that it is difficult, and perhaps impossible, to account for the halo field stars of the Milky Way exclusively by disrupted globular clusters. However, there seems little doubt that dynamical destruction does remove some clusters from the initial population in the Milky Way and other galaxies. For example, even if we take the metallicity and abundance constraints at face value, and assume that the halo field stars formed independently from the globulars, the present-day excess of field stars to stars in globulars (roughly

100 to 1) suggests that many globular clusters could have been disrupted without their stars 'contaminating' the field star population to an observationally measurable degree. Thus it is certainly possible that the original number of globular clusters in the Milky Way was significantly greater than it is today. Clearly, the detection of CN-rich stars in the Milky Way halo would be an important step in detecting disrupted globular clusters, although the absence of such stars only requires that disruption occur before CN enhancement.

A related issue is the possibility that the bulge of the Milky Way and the nuclear regions of other galaxies may have been produced, at least in part, by disrupted globular clusters (e.g. Aguilar *et al.* 1988). This is theoretically plausible since the higher stellar densities in such regions and shorter orbital periods for globular clusters at small galactocentric distances make destructive processes more effective. The presence of a galactic bar and globular clusters on box orbits also enhance the destruction rate (Long *et al.* 1992; Ostriker *et al.* 1989). However, observations suggest that disrupted globular clusters do not contribute significant stellar mass to galactic bulges. The two principle problems are the higher typical metallicities of bulges relative to globular clusters (Surdin and Charikov 1977; van den Bergh 1991b) and the lack of preferential removal of the most massive clusters through dynamical friction in the central regions of galaxies. The latter process would be apparent through a variation of the GCLF with radius in the inner regions of galaxies, which is not observed (see Chapter 5). McLaughlin (1995) notes for the specific case of M87 that if the cluster distribution initially followed the light distribution of the galaxy and if dynamical destruction is responsible for the much larger core radius of the globular cluster system observed at the present epoch, then at most 30% of the nucleus of M87 could have been produced by disrupted globular clusters.

The extension of globular cluster destruction studies to galaxies other than the Milky Way is gaining momentum, thanks in part to the more detailed information from recent observations available for these systems. A theoretical study by Murali and Weinberg (1996) concentrates on how destruction rates depend on the properties of the host galaxy, thereby opening a new avenue for empirical tests of the importance of destruction processes. A key result from this work is the prediction that globular cluster destruction is more efficient in lower-luminosity elliptical galaxies. This is a consequence of galaxy central surface brightness (or stellar density) increasing with decreasing integrated galaxy luminosity. Bulge-shocking is therefore more effective in lower-luminosity ellipticals than in the brightest ellipticals. Consequently, some part of the increase of S_N with elliptical galaxy luminosity may be a result of more efficient globular cluster destruction in lower luminosity galaxies. However, the models of Murali and Weinberg (1996) appear to be unable to account for the entire observed effect. Destruction mechanisms are also unable to provide an explanation for the variation of S_N at a given galaxy luminosity, including the particularly high S_N of some, but not all, highly luminous ellipticals at the centers of groups and clusters. (Interestingly, bulge-shocks are not expected in dwarf *spheroidal* galaxies, which often have high values of S_N.)

In discussing globular cluster destruction, it is important to draw a distinction between long-term dynamical processes and disruption on much shorter timescales which can result from stellar evolution. In the latter case, the orginal globular cluster population may have been much bigger than the one observed today, but most of the clusters were destroyed by stellar mass-loss and/or supernova explosions which occurred when such clusters were in their infancy. In this case, it is questionable whether the disrupted clusters warrant the term 'globular', since one could argue that they were doomed to be destroyed from the outset. We discussed this issue in the previous chapter in the context of the young globular cluster candidates found in merging and interacting galaxies.

Another physical process which can constrain the masses and radii of globular clusters is related to tidal limitation, described in Chapter 3, in which a cluster must have a higher mean density than the surrounding environment (see equation (3.1)). This is likely to be equally valid for protoglobular gas clouds. It is hard to see how such clouds could form and fragment into stars if they did not have a mean density higher than their surroundings. It seems probable that this effect limits the initial distribution of globular cluster properties, preventing low-density systems from forming, but, as noted in Chapter 3, the absence of high-density globulars at large distances from the Galactic center cannot be explained by this process.

The above discussion suggests that there is still room for the idea that the globular clusters observed today are only a small fraction of the initial globular cluster population. However, the difference between halo field stars and globular clusters seems to suggest that if significant globular cluster disruption occurred the timescale was relatively short. Further, there may well have been a more numerous population of marginally bound or unbound star clusters that formed along with the genuine globulars, but these objects were destined to be destroyed on short timescales through stellar evolutionary processes. These objects may have been more similar to the open clusters of the Milky Way. Indeed, as discussed in Chapter 5, observations of candidate young globular cluster systems require that considerable globular cluster destruction takes place if these systems are to resemble old globular cluster systems in a Hubble time.

7.2   Primary formation models

In an attempt to impose some order on the large number of models of globular cluster formation, we follow Fall and Rees (1988) and separate the models into three classes. Primary formation models are those in which globular clusters form before galaxies; secondary models have galaxy and globular cluster formation occurring roughly contemperaneously; and tertiary models are those in which globular clusters form after their host galaxies. The distinction between secondary and tertiary models is complicated by the fact that galaxy formation is probably not a tidy event (e.g. Peebles 1989). In many galaxy formation models, the stellar component of a galaxy builds up over time, through processes such as gradual accretion, hierarchical clustering, and mergers (Silk and Wyse 1993 and references therein). Thus the point at which a galaxy 'forms' is to some extent a semantic question. Here, we regard globular cluster models as tertiary if clusters form after the host galaxy has settled down to a relatively quiescent state. We define secondary models as those in which the bulk of the globular clusters around a galaxy form when the galaxy itself is still in the process of protogalactic collapse or is undergoing significant evolution.

If globular clusters are exclusively ancient objects, one possibility is that they all form through some mechanism that is peculiar to the early universe. Such a scenario was proposed by Peebles and Dicke (1968) who noted that in hierarchical clustering models of structure formation the Jeans mass just after recombination is around 10^6 $M_\odot$. The first bound objects to form in such models therefore have masses similar to those of globular clusters. Subsequently, it has been shown that this mass-scale depends on both cosmological and astrophysical factors (e.g. Gunn 1980; Carr and Rees 1984; Ashman and Carr 1988, 1991), but a mass-scale comparable to that of globular clusters does arise fairly naturally.

As described throughout Chapter 5, the spatial distribution, number, and metallicity of globular clusters depends on properties of the galaxy with which it is associated. It is difficult to imagine how this could be set up if the globular clusters formed before and independently of galaxies. It is possible to account in some way for the spatial distribution and number through 'biasing' globular cluster formation, such that globulars only form from high-density peaks in the matter distribution. Rosenblatt *et al.* (1988) introduced this idea in the contest of hierachical clustering scenarios such as the cold dark matter model (Peebles 1984; Blumenthal *et al.* 1984), in which the stellar component of a globular cluster forms within a halo of non-baryonic dark matter.

This idea has been generalized to other hierarchical clustering models by West (1993), who notes that such peaks are more common when the local environment itself is of high-density. Further, elliptical galaxies are found preferentially in high-density environments. West (1993) shows that the higher specific frequency of globular clusters around ellipticals relative to spirals can be explained by the biased globular cluster formation scenario. The possibility that specific frequency increases with environmental density for a given morpholgical type of galaxy is also explained, although, as discussed in Chapter 5, evidence for such a trend is not conclusive.

The main advantage of primary formation models is that they naturally explain why the globular cluster luminosity function is independent of galaxy type or environment. However, it seems that only the biased version can account for the steep density profiles of globular cluster systems (Rosenblatt *et al.* 1988), which presented a problem for the original Peebles and Dicke (1968) proposal (see also Gunn 1980).

There are, however, major drawbacks to primordial formation models. Probably the most serious problem arises from the evidence that globular clusters do *not* have dark matter halos (see Chapter 2). All primordial models that we are aware of that are also consistent with constraints on cosmological models require globular clusters to have such halos. By itself, this appears to rule out primordial models. Biased globular cluster formation addresses the issue of the correlations between the number and spatial distribution of globulars with properties of the host galaxy, but seems unable to account for metallicity gradients. A related issue is the presence of the disk globular clusters in the Milky Way. A flattened, rotating population of exclusively metal-rich globulars seems impossible to reconcile with primordial models. It is conceivable that halo clusters are primordial and that the disk clusters have some other origin. However, the luminosity functions of the disk and halo clusters are indistinguishable. This effectively nullifies the main attraction of primordial models of naturally explaining the universality of the globular cluster luminosity function. Young globular clusters in mergers present a similar problem. Again, such objects are clearly not primordial.

On the basis of these considerations, we feel that primordial globular cluster formation models are inconsistent with current observational data. It is conceivable that the high S_N values of some cDs may result from 'excess' globular clusters associated with the galaxy cluster potential, in which case one possible origin for these clusters is primordial formation (see West *et al.* 1995). However, this requires that the excess clusters have a different origin than the majority of globular clusters which are clearly associated with individual galaxies.

7.3 Secondary formation models

The majority of globular cluster formation models assume that most globular clusters formed at roughly the same time as the host galaxy. This general picture is attractive because of the correlation between some globular cluster properties and position in the parent galaxy (notably the metallicity gradients found in the globular cluster systems of ellipticals discussed in Chapter 5), and because the ages of globular clusters suggest they formed at an epoch when their parent galaxies formed.

No current formation model addresses the origin of all the properties of globular clusters in detail. The trend in the literature has been to start with one particular aspect of globular clusters (such as mass or metallicity homogeneity) and to develop a formation theory from there. We use this difference in starting point as a means of classifying secondary models. This approach allows us to emphasize the common themes that appear in formation models. It seems likely that a complete model of globular cluster formation will draw on ideas based on more than one of these property-specific models.

7.3.1 *Globular cluster mass and the mass distribution*

The globular cluster luminosity distribution, when plotted in terms of magnitude, shows a distinct peak (see Chapter 3). This has led to the idea that globular clusters have a characteristic mass which, perhaps significantly, is greater than the mass of other stellar clusters and associations. Several models of globular cluster formation are largely motivated by an attempt to explain the origin of this mass-scale.

One influential scenario of this kind was proposed by Fall and Rees (1985). These authors suggested that thermal instability in the gas of a protogalaxy would lead to a population of cool clouds embedded in a surrounding medium of hotter, more tenuous gas. The cool clouds are assumed to be in pressure equilibrium with the surrounding medium. In clouds with a low metal abundance, cooling becomes relatively inefficient below about 10^4 K, so that the clouds are supposed to 'hang up' at this temperature for a significant time. Fall and Rees (1985) noted that the Jeans mass of such clouds is of order 10^6 M$_\odot$ and that only clouds above this mass scale are likely to fragment into stars. Since this mass is similar to that of globular clusters, Fall and Rees (1985) identified the cool clouds with the progenitors of globular clusters.

Some features of this picture are reminiscent of the earlier ideas of Gunn (1980) and McCrea (1982), who invoked strong shocks within protogalactic gas to imprint the globular cluster mass scale. Such shocks increase the density of the gas locally, thereby increasing the cooling rate. This results in dense gas clouds with temperatures around 10^4 K, as is achieved in the Fall–Rees picture through thermal instability.

The Jeans mass of clouds arising from shocks is also similar to that in the Fall–Rees picture.

An important difference between Gunn's (1980) discussion and that of Fall and Rees (1985) is the length of time the clouds remain at a temperature of 10^4 K. The Fall–Rees picture specifically requires that the cooling time at this temperature is appreciably longer than the dynamical time of the clouds. They claim that this is required to imprint a mass-scale of 10^6 $M_\odot$ on the resulting population of clusters. Gunn (1980), on the other hand, requires sufficient cooling at this temperature that the cooling time is *less* than the dynamical time. He argues that only in this way can the clouds fragment to stars, and further suggests that if cooling is inefficient at 10^4 K the gas cloud is more likely to form a single supermassive star! This divergence of theoretical opinion from similar starting points illustrates one of the hazards of constructing globular cluster formation models.

Interestingly, the requirement by Fall and Rees (1985) that cooling virtually stops at 10^4 K has proved to be a problem for the model. Even in a gas that contains no metals, cooling can proceed below this temperature in a variety of ways. The most important process is non-equilibrium cooling, which can generate significant amounts of H_2 (MacLow and Shull 1986; Shapiro and Kang 1987; Haiman *et al.* 1996). The presence of H_2 drastically reduces the cooling time at 10^4 K, potentially allowing the clouds in the Fall–Rees picture to cool to around 10^2 K. The situation is even worse if there are any metals present in the gas, since this will again reduce the cooling time.

These drawbacks were addressed by Kang *et al.* (1990), who argued that the characteristic mass of the Fall–Rees picture could be preserved, provided there was a sufficient flux of UV or X-ray radiation at the time of globular cluster formation. Under certain conditions, such a flux can prevent the clouds cooling below 10^4 K. While this may alleviate the problem, Ashman (1990) pointed out that there were difficulties with the two sources of radiation suggested by Kang *et al.* (1990). If an early generation of massive stars produced the radiation over the required time period, the metals produced in these stars would probably pollute the Milky Way to metallicity levels much larger than observed. The other possibility is radiation from active galactic nuclei, but the energy requirements would lead to more background radiation than is observed (Cavaliere and Padovani 1989; Ashman 1990).

Other problems are reminscent of those facing primordial formation models and suggest that, at best, the Fall–Rees picture can only account for *some* globular clusters. The high metallicity of the disk globular clusters of the Milky Way (and higher-metallicity clusters around many ellipticals) almost certainly formed out of gas that experienced significant pre-enrichment. In such gas, for the reasons outlined above, there is nothing 'special' about a temperature of 10^4 K and thus no characteristic mass in the Fall–Rees sense. Many dwarf galaxies such as Fornax have globular clusters which are essentially indistinguishable from those around massive galaxies (see Chapter 4), but the virial temperature of such galaxies is too low to have ever supported the thermal instability required in the Fall–Rees picture. If the candidate young globular clusters in the Large Magellanic Clouds are genuine globulars, they did not form in the manner envisaged in the Fall–Rees picture.

The Fall–Rees scenario has a couple of additional minor problems. First, the majority of Milky Way globular clusters have masses well below the predicted minimum mass. This problem may be alleviated by allowing for the uncertainties in the model and the possibility of mass loss from globular clusters. Second, the minimum mass of globular clusters is expected to increase with Galactocentric distance, with $M_{min} \propto R^{1/2}$. This is a general prediction for galaxies with roughly isothermal dark halos. Such a trend is not observed in the Milky Way and other galaxies.

The conditions of the gas in the Fall–Rees scenario are similar to those in 'cooling flows' observed primarily in the central galaxies of galaxy clusters (see Fabian 1994 for a review). There has been some speculation that similar cooling flows on galactic scales might be responsible for globular cluster formation (Fabian *et al.* 1984; Ashman 1990). However, the finding that the mass flow rate in galaxy cooling flows at the present epoch is uncorrelated with the specific frequency of globulars in the host galaxy, along with other more detailed problems, strongly suggests that present-day cooling flows are not responsible for any significant formation of globular clusters (e.g. White 1987; Harris 1991; Bridges *et al.* 1996). Whether one regards certain kinds of organized protogalactic collapse as analogous to cooling flows seems to be a matter of taste, so that in this limited context it may be possible to associate the cooling flow phenomenon with globular cluster formation.

Despite difficulties with the details of the Fall–Rees scenario, some of its key ingredients have been used in other models. Most notably, other pictures have incorporated the idea that protoglobular clouds exist in pressure equilibrium with a surounding hot gas. Murray and Lin (1992) present an argument which they claim provides empirical evidence for pressure-confined protoglobular clouds. By using the observed properties of Milky Way globulars, along with certain assumptions about the evolution of the gas before star formation occurs in the protoglobular clouds, they derive various properties of the clouds. These authors claim that the pressure in protoglobular clouds decreases with increasing Galactocentric distance, and that the trend roughly follows the expected pressure profile of the hot gas in which the clouds are embedded. This decrease in 'pressure' is essentially another way of describing the increase in globular cluster half-light radius with Galactocentric distance described in Chapter 3.

It is not apparent to us that Murray and Lin's (1992) interpretation of the data is unique. In general, the low density of remote globular clusters seems just as likely to stem from the decrease in the Galactic potential with radius as from the decrease in pressure of the hypothetical hot protogalactic gas. Specifically, we argued in Section 7.1 above that tidal limitation of protoglobular gas clouds may explain the increase in $r_{1/2}$ with R. Murray and Lin (1992) suggest this is unlikely on the grounds that current globular clusters sit well above the tidal limitation criterion. However, unless globular clusters have primarily circular orbits, most of them are currently located at Galactocentric distances far beyond their closest approach to the Galactic center. It is therefore expected that their densities should be somewhat higher than the tidal limit at their current location.

To quantify this idea, we note that Peterson (1974) finds a mean eccentricity of globular cluster orbits of 0.5, implying a distance at perigalacticon around one-third that at apogalacticon (de Zeeuw 1985; Okazaki and Tosa 1995). Since globular clusters are more likely to be observed when near apogalacticon, we approximate the perigalacticon distance as $R_p = R/3$, where R is the currently observed Galactocentric

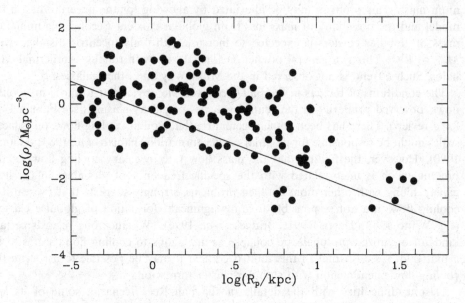

Fig. 7.1. The density of globular clusters against perigalactic distance. The solid line represents the tidal limitation criterion.

distance of a cluster. In Figure 7.1 we plot mean globular cluster density within r_t against R_p. The solid line represents the tidal limit due to the potential of the Milky Way. This is derived using equation (3.3) and approximating the mass distribution of the Milky Way as a singular isothermal sphere with an associated rotation velocity of 220 km/s. There is clearly considerable scatter at all radii, and some clusters actually fall *below* the tidal limit, but the overall impression is that the clusters tend to follow the tidal limitation criterion. It is possible that orbital information derived from proper motion studies will enable a better estimation of perigalactic distances. If the use of such distances reduces the scatter in Figure 7.1, it will provide some evidence that the observed densities of globular clusters are determined, at least to some extent, by the tidal limitation criterion.

Like the Fall–Rees picture, the scenario proposed by Murray and Lin (1992) requires that the embedded protoglobular clouds are heated by some external source, but the motivation is somewhat different to that of Fall and Rees (1985). Using their derived temperatures and densities of protoglobular clouds, Murray and Lin (1992) find that the clouds had cooling times significantly shorter than their dynamical times. These authors suggest that it is impossible to form clouds with such properties since they are not initially in equilibrium. Their solution is to suppose that external heating allows thermal equilibrium to be maintained. We note that the short cooling times are only a problem if the only source of pressure support in the clouds is thermal: turbulence or magnetic pressure can keep such clouds in equilibrium even if the cooling time is shorter than the dynamical time. Indeed, both processes are probably important in providing support within giant molecular clouds (e.g. McKee *et al.* 1993 and references therein; Harris and Pudritz 1994).

The equilibrium temperature of clouds found by Murray and Lin (1992) is around 10^3 K, an order of magnitude lower than in the Fall–Rees picture. The energy requirements to maintain this temperature through external heating are consequently more modest. Further, the Jeans mass in this picture is interpreted as an *upper* mass to the globular clusters that eventually form with a value around 10^5 $M_\odot$. The subsequent evolution of these clouds, including fragmentation and ensuing star formation, has been investigated by various authors including Lin and Murray (1992). We shall have more to say about this in the next section. Another feature of the Murray and Lin (1992) proposal is that they favor cloud–cloud collisions for triggering star formation within the protoglobular clouds. This is a theme which appears in other models described below.

While the picture advocated by Murray and Lin (1992) overcomes some of the difficulties of the Fall–Rees picture, the reliance on a two-phase protogalactic medium leaves it vulnerable to the same problems outlined earlier. It is also worth noting that, like the Fall–Rees model, the Murray and Lin (1992) scenario leads to the expectation that globular cluster mass increases with Galactocentric distance, contrary to observation. The form of this increase depends primarily on the density profile of the dark matter halo of the Galaxy, but in the case considered by Murray and Lin (1992) the dependence is linear with Galactocentric distance. The picture predicts a similar dependence for other galaxies which, as noted above, conflicts with observation.

A more general problem for any model that relies on a cooling condition or a heating–cooling balance to preserve a special temperature, and thus a characteristic Jeans mass, is that more metals in the gas inevitably lead to more efficient cooling. Qualitatively, one expects all such models to lead to the prediction of a decrease in globular cluster mass with increasing metallicity. For example, Richtler and Fichtner (1993) propose a Murray–Lin-type scenario in which only low-metallicity clouds give rise to globular clusters. They support their scenario by noting that young globular cluster candidates in the Magellanic Clouds tend to have low metallicities. However, the mounting evidence of old, high-metallicity clusters in elliptical galaxies probably presents a fatal problem for this and related models.

One further theoretical complication with the two-phase medium of cool clouds and hot gas is exactly how the clouds cooled in the first place. For the mass-scales associated with protoglobular clouds, cooling from the virial temperature of a bright galaxy (around 10^6 K) to 10^4 K or below may include an isochoric (constant volume) phase, followed by a strong shock traveling into the cloud (e.g. Ashman 1990). This complication is embraced by Vietri and Pesce (1995). In their model, low-mass clouds are completely reheated by the bouncing shock, whereas in high-mass clouds the shocked material forms a dense shell and fragments into an unbound system of stars (identified with the field stars of the Galactic halo). In clouds of intermediate mass the bounced (outward-moving) shock decelerates the inward-moving dense shell of material *before* it has fragmented, leading to the formation of a cold, dense cloud in pressure equilibrium with the hot gas in the protogalaxy. These clouds are supposed to fragment into globular clusters. Since high-mass and low-mass clouds have different fates, the scenario leads to a predicted mass *range* of globulars which is roughly consistent with observation. Further, the lower mass limit is insensitive to the metallicity of the gas. Unfortunately, the upper mass is reduced by an order of magnitude once the metallicity exceeds [Fe/H] ~ -2. This dependence of mass on metallicity is a serious drawback.

Perhaps the most notable aspect of the models discussed above is that they bear little relation to formation models of other types of star cluster. The view that globular clusters are fundamentally different to other star clusters probably led to this bias. This is a sensible position: globular clusters *do* differ from other clusters. However, this does not rule out the possibility that all star clusters form in a similar manner and *globular* clusters are those objects with properties that allow them to survive for a Hubble time.

The philosophy that globular clusters represent a distinct class of objects has tended to divorce models of globular cluster formation from observations of star cluster formation. Larson (1996, 1993, 1990 and earlier papers) has taken a different approach, emphasizing what can be learned about globular cluster formation from studies of star-forming regions. In these papers, Larson has noted that *bound* star clusters only apppear to form in the dense cores of much larger star-forming clouds. The fact that globular clusters are bound sets them apart from many open clusters which disperse on a relatively short timescale (see Richtler and Fichtner 1993), so it seems plausible that globular clusters may form from similar dense cores. If this is the case, then, by analogy with current bound cluster formation, one expects the total mass of the protoglobular cloud to be two to three orders of magnitude greater than the current mass of globular clusters. In other words, the formation efficiency of *bound* clusters in such environments is between 10^{-2} and 10^{-3}. It is perhaps suggestive that the fraction of stars in globular clusters relative to field stars in the Milky Way halo is around 1% (e.g. Ashman and Zepf 1992). Further, the highest S_N values observed in some dwarf spheroidals and giant ellipticals correspond to a globular cluster fraction approaching 1% of the total stellar population.

The idea that protoglobular clouds have masses around 10^8 $M_\odot$ and that only the cores of these clouds survive as bound globular clusters is very different to the secondary formation models described above. Further, it alters the fundamental question of why globular clusters only form at certain times or in certain environments. The central issue to understanding globular cluster formation becomes the question of why and how such massive gas clouds form. Several possibilities have been proposed in the literature (see Larson 1993 for a summary), some of which we discuss below.

The possibility that globular cluster formation is analagous to the formation of other star clusters can be used to develop primary, secondary and tertiary formation models. However, it has been most extensively discussed in the context of secondary models, hence its inclusion in this section. The work of Harris and Pudritz (1994) is notable in this regard, since it relies heavily on observations of giant molecular clouds (GMCs) to construct a globular cluster formation model.

As described in Chapter 3, the Milky Way globular cluster luminosity (or mass) function can be fit by a power law of the form $N_{GC} \propto M^{-\alpha}$. This power law holds above a minimum mass M_{min}; below this mass the power law flattens off. The value of α is sensitive to the choice of M_{min}, but for the Milky Way globular clusters, Harris and Pudritz (1994) adopt $\alpha = 1.7$ from Secker (1992), which assumes $M_{min} = 7.5 \times 10^4$ $M_\odot$. As described in Chapters 4 and 5, a similar exponent fits the globular cluster mass functions of M31 and several elliptical and cD galaxies where suitable data exist (Harris and Pudritz 1994). (McLaughlin (1994) suggests that the value of α may be a little higher in ellipticals relative to spirals.) For the populous systems of ellipticals and cDs, the power law breaks down at both low and high masses, with the exponent steepening above about 3×10^6 $M_\odot$. One would not expect to see such a break in the Milky Way and M31 because of the small number of clusters at such high masses.

Harris and Pudritz (1994) make the important observation that the value of α for globular cluster systems is similar to that of the mass spectrum of GMCs, the cores of GMCs, and HII regions. These authors conclude that this apparent universality of α argues that globular cluster formation occurred in a similar manner to star cluster formation observed today in the Milky Way. Their model of globular cluster formation therefore exploits observations and theoretical considerations relating to the formation of 'normal' star clusters. The starting point is the idea that the required mass spectrum can be produced by cloud–cloud coalescence. Various studies have shown that such a process can produce a mass spectrum with α in the required range (e.g. Field and Saslaw 1965; Kwan 1979; Kwan and Valdes 1983; Carlberg and Pudritz 1990).

Like Larson (1993 and earlier papers), Harris and Pudritz (1994) conclude that proto-globular clouds, which they call 'super GMCs', must have masses around 10^8 $M_\odot$: three orders of magnitude greater than GMCs found in the Milky Way today. The physical conditions in these super GMCs have been further explored by McLaughlin and Pudritz (1996). As in the formation models described above, Harris and Pudritz (1994) require that the super GMCs remain stable for an appreciable amount of time. This is not specifically to ensure that the globular cluster mass-scale is imprinted, as in the Fall–Rees scenario and related pictures, but is required in order for sufficiently massive clouds to form through agglomeration. Rather than relying on heating to prevent cloud fragmentation, Harris and Pudritz (1994) suggest that magnetic fields and turbulence provide the necessary support, as is the case in the GMCs of the Milky Way.

Note that the mass scale of these super-GMCs is similar to that of the gas fragments of the Searle and Zinn (1978) picture discussed in Chapter 3. Such fragments are sometimes identified with satellite dwarf galaxies (Larson 1993, and references therein). The primary difference between the Searle–Zinn picture and the Harris and Pudritz (1994) scenario is that the latter requires the super-GMCs to be built up from smaller clouds within the protogalaxy.

While it does not appear to be a critical aspect of the model, Harris and Pudritz (1994) note that if super GMCs are self-gravitating *and* pressure-confined by surrounding hot gas, the increase in $r_{1/2}$ with Galactocentric distance is explained (see Murray and Lin 1992 and above). They do *not*, however, envisage super GMCs condensing out of such a hot gas by thermal instability, thereby avoiding the difficulties inherent in models which incorporate this idea.

The principal criticism of the super-GMC model is that there is no evidence that such massive molecular clouds form in practice. If one is simply explaining the formation of old globular clusters, it is clearly possible (although *ad hoc*) to postulate that conditions in the early universe were conducive to the formation of larger molecular clouds. The situation is more problematic if one regards the young globular candidates described in Chapter 5 to be genuine analogs of old globulars. In merging and interacting galaxies, it seems the only candidates for protoglobular gas clouds are the GMCs found in disk galaxies. Merging and interaction may allow such clouds to coalesce and form more massive clouds, but the timescales involved suggest that it is difficult to form clouds of masses around 10^8 M$_\odot$ required by the Harris and Pudritz (1994) model. Other problems include the low metallicity of the super-GMCs (necessary to form metal-poor globular clusters), which suggests that the gas is more likely to be atomic than molecular.

Another possible starting point for globular cluster formation models that retains the link to observed star cluster formation is to suppose that star formation *efficiency* is somehow enhanced in protoglobular clouds. This has the attraction of invoking ordinary GMCs as the progenitors of globular clusters at the price of requiring an increase in the star forming efficiency of at least two orders of magnitude. This route has been followed by Jog and Solomon (1992), who suggest that galaxy collisions ionize HI clouds in disk galaxies and that the high pressure in this gas sends a shock into pre-existing GMCs. In this way, the star formation efficiency may be increased. The necessary star formation efficiencies are quite intimidating (greater than about 50% compared to typical values in GMCs of less than 1%) and it seems plausible that *both* larger GMC masses and higher star formation efficiencies may be involved in the formation of globular clusters.

A recent model advanced by Elmegreen and Efremov (1997) also makes a direct attempt to link globular cluster formation to the formation of other types of star cluster. Like Harris and Pudritz (1994), these authors note the similarity in the exponent of the power-law mass spectrum of star clusters and argue that it is evidence of a universal formation process for all star clusters. In their model, the origin of the mass spectrum stems from the fractal nature of turbulent gas. Structural differences between the different types of star cluster arise from differences in pressure at the time of formation, with globular clusters forming from regions of higher pressure than open clusters. As is the case in the Harris and Pudritz (1994) model, Elmegreen and Efremov (1997) require that low-mass clusters are preferentially destroyed in order to reproduce the observed mass distribution of old globular clusters in the Milky Way and other galaxies.

7.3.2   Self-enrichment or pre-enrichment?

A second starting point for models of globular cluster formation is the remarkable star-to-star homogeneity in iron-peak elements observed within individual globular clusters. There are two general approaches for explaining this homogeneity. Some models assume that the metals in globular clusters were formed before the clusters themselves. In models that incorporate such pre-enrichment, the mechanism for metal generation need not be linked to the process of globular cluster formation itself. The assumption is simply that an earlier generation of stars enriched the gas out of which globular clusters form. However, pre-enrichment *does* require that the gas in protoglobular clouds is well-mixed, washing out any spatial variations that might have initially existed, and that the epoch of star formation is brief. Rapid star formation ensures that metals produced in the globular cluster stars do not have time to pollute stars forming later within the cluster, so that all stars in an individual globular cluster are required to be coeval.

The alternative point of view is that protoglobular clouds are comprised of primordial, metal-free gas, so that their observed metallicities are due to self-enrichment. Generically, this demands that the first burst of stars in globular clusters is no longer visible and that some mechanism ensures that the metals are well-mixed before the bulk of the globular cluster stars form. The first condition is rather easy to meet: provided the first stars are sufficiently massive, they will now exist only as stellar remnants within globular clusters. Since it is precisely the massive stars which produce metals on a short timescale, this condition is naturally satisfied, although it does require that *only* massive stars make up this first generation.

The second problem of mixing the metals is more difficult to deal with and is usually regarded as the main problem facing self-enrichment models. Upper main sequence stars are required to produce the observed metals, but the lifetime of such stars (around 3×10^6 years) is comparable to the dynamical timescale of globular clusters. The timescales for star formation throughout the protoglobular gas cloud and for diffusion of metals through the cloud are likely to be appreciably longer than the dynamical time. Thus it is expected that stars forming in different parts of the protoglobular cloud will form out of gas enriched to different levels, contrary to observation.

Before discussing the various ways in which self-enrichment models circumvent this problem, we note that a similar difficulty is also faced in pre-enrichment models. In such pictures, the gas may start out well mixed, but star formation still needs to occur on a relatively rapid timescale to prevent recyling and an increase in the metal content of the gas in the still-forming cluster. One option in pre-enrichment schemes is to suppose that there are no massive stars to produce metals.

It is also worth noting that some globular clusters must have experienced a degree of pre-enrichment. It seems inevitable that the metal-rich disk globulars of the Milky Way and some globulars in ellipticals formed out of pre-enriched gas. Thus the challenge for self-enrichment models is to explain the metallicities of globular clusters that are more likely to be 'first generation'. The best-studied candidates for such objects are the halo globular clusters of the Milky Way, so that the observed metallicities of these objects provide the most useful constraints on enrichment models.

One important clue to the enrichment problem is the observation that the clearest case of iron-peak metallicity variations within a Milky Way globular occurs in ω Cen (see Section 2.3). This object is the most luminous of the Milky Way globulars. M22 *may* have similar variations (Section 2.3) and is the third most luminous globular cluster in the Milky Way. There are at least two (possibly related) interpretations of this finding. First, it suggests that only the most massive clusters retain appreciable amounts of gas, so that when supernovae explode there is still material out of which additional stars can form. A second intepretation is that many young globular clusters may retain gas, but that in clusters which suffer supernova explosions, only the most massive can survive the event. The idea that many young clusters are disrupted by such explosions is a feature of several models, as described below.

An ingenious approach to the problems inherent in self-enrichment models was devised by Cayrel (1986). The model assumes that a protoglobular cloud is characterized by a dense central core and a surrounding infalling envelope of material. Such a situation has been found to arise in spherical collapse models of gas clouds (e.g. Larson 1978). According to Cayrel (1986), the high central density means that star formation occurs first in the core, and that the first stars to form there are massive. These massive stars produce supernovae which drive material outward. The interaction between this outflow and the inflowing envelope of material produces a shock where the bulk of the star formation is supposed to occur. Since the material expelled from the supernovae is metal-enriched, the stars formed in the globular cluster that are still observable today have non-zero metallicity, even if the original gas cloud was metal-free. While Cayrel (1986) does not investigate the question in detail, he speculates that turbulence in the shock ensures that metals are well-mixed. He also notes that the enhanced [O/Fe] ratios relative to the solar neighborhood found in globular clusters are consistent with metal-enrichment through supernovae deriving from massive stars. As we discuss below, both basic ideas have been exploited in subsequent self-enrichment models.

A similar picture was proposed by Dopita and Smith (1986), who also addressed the question of the mass of the cloud required to contain the supernovae ejecta. These authors were primarily concerned with explaining the metallicity *inhomogeneities* in ω Cen, assuming that such inhomogeneities result from supernovae. Dopita and Smith (1986) found that the abundance variations in ω Cen require multiple supernovae events, and that the mass of the gas cloud out of which the cluster formed had a mass around [Fe/H] $\simeq -2$ to -1. More generally, they found that only clouds with masses greater than about 10^6 $M_\odot$ survive supernova explosions at all, the most disruptive effect being the momentum of shells swept up by supernova explosions.

Morgan and Lake (1989) also looked at the same basic scenario, with the improvement of a more accurate cooling function and a specific density profile in the gas cloud. These authors focussed on the minimum mass of globular clusters that could be self-enriched by supernovae and survive disruption, arriving at a value of $10^{4.6}$ $M_\odot$. There are Milky Way globular clusters with masses below this value, but the minimum mass is clearly in the right ballpark.

Brown *et al.* (1991) advance a scenario similar to that of Cayrel (1986) in which stars currently observed in globular clusters form in a shell previously enriched by supernovae. Metallicity homogeneity is assumed to arise through efficient mixing in the shell of swept-up material. A limit on the degree of enrichment is obtained using dynamical constraints related to the potentially disruptive effects of the supernovae. Brown *et al.* (1991) derive a metallicity range for first generation globular clusters of around 10^{-2} to 10^{-1} $Z_\odot$, consistent with observation.

The efficient mixing of metals in swept-up shells of material is a feature of many self-enrichment models. A detailed study of the problem by Murray and Lin (1990) suggests that this may not arise in practice, since the timescale for mixing processes is rather long. Another, somewhat unattractive, aspect of all self-enrichment models is the fact that the stars in globular clusters observed today are necessarily a second generation of stars, the first having produced the metals. As noted above, metal-producing stars themselves are only likely to exist as stellar remnants today and consequently do not present a problem, but such scenarios demand that very few lower-mass stars formed at the same time as the massive stars. Such a top-heavy initial mass function tends to compound the problem of keeping the clusters bound.

7.4 Tertiary formation models

Several scenarios have been proposed which, to varying degrees, incorporate the idea that globular clusters form after their host galaxy. As noted above, there is overlap between some of these pictures and secondary formation models, so that in some cases the distinction becomes largely semantic. Part of the reason for this overlap stems from the popular hierarchical clustering models of galaxy formation in which protogalaxies are lumpy and assemble over an appreciable period of time.

The disk globular clusters of the Milky Way provide smoking gun evidence that at least some globular cluster formation is tertiary. It seems inevitable that the Milky Way must have completed its early formation before these disk clusters formed, otherwise it is difficult to see how such clusters 'knew' to form in the plane of the disk. Further, most galaxy formation models, backed by various lines of evidence, suggest that much of the Galactic spheroid was assembled before the disk. Formation models for disk globulars tend to mirror formation models of the thick disk itself. If the thick disk is just an intermediate stage in the collapse process between spheroid and thin disk, it is natural to regard globular cluster formation as occurring when the thick disk is still in a gaseous phase (e.g. Burkert *et al.* 1992). In this case, many of the secondary formation models described above are applicable to this specific situation, provided, of course, that they can explain the formation of high-metallicity globular clusters. Other pictures are more intimately linked to the expected properties of the (gaseous) thick disk. Murray and Lin (1991) invoke compression of gas clouds crossing the still gaseous disk to produce the disk population of clusters. Ashman and Zepf (1992) suggest that a minor merger between a satellite galaxy and the young Milky Way disk gives rise to these globular clusters, reflecting suggestions that the stellar thick disk is produced in a similar manner. In this picture, globular clusters may either form from gas in the young Milky Way disk, or gas associated with the infalling satellite galaxy.

The evidence for distinct epochs of globular cluster formation in some elliptical galaxies discussed in Chapter 5 also suggests that some globular clusters form after the bulk of the host galaxy has settled down. In the merger picture, the situation again tends to reduce to one of semantics. The metal-rich globulars formed in the merger clearly formed after the progenitor spiral galaxies, but contemporaneously with the elliptical galaxy itself. However, attempts to explain multiple globular cluster populations in the context of monolithic collapse models presumably need to invoke globular cluster formation initially switching off with one or more additional formation episodes occurring at later times. If such models can be constructed, some globular cluster formation would necessarily be tertiary.

Rodgers and Paltoglou (1984) and Larson (1986, 1996) have raised the possibility that all globular clusters may form in small disk galaxies, so that globular clusters in spheroidal distributions such as those of the Milky Way halo arrived at their present location via hierarchical clustering and/or merging. Part of the motivation for this idea stems from the observation that, at the present epoch, it is primarily disk galaxies that give rise to significant star formation. Further, the brightest young star clusters in small disk galaxies like the Large Magellanic Cloud are regarded by some workers as genuine young globulars. In terms of galaxy formation models, this idea has the attraction of fitting naturally with current models in which all galaxies are built from smaller subunits. However, the physics of globular cluster formation is not addressed in detail in this picture at present.

Perhaps the most distinct tertiary formation model is the proposal that many globular clusters are the surviving cores of nucleated dwarf ellipticals (Zinnecker 1988; Freeman 1990). This suggestion was motivated by the observation that the nuclei of these galaxies have certain characteristics in common with globular clusters, and that such galaxies are common in galaxy clusters where the most populous globular cluster systems are found. The idea is that parent galaxies accrete these dwarf galaxies, stripping away the outer stars so that only the dense nuclei survive. The main stumbling blocks to this picture include the observation that, in detail, the cores of nucleated dwarfs tend to be significantly more massive than most globulars (e.g. Meurer *et al.* 1992) and that correlations between globular cluster properties and location in the host galaxy are difficult to understand in any picture where globulars are accreted (cf. the comments on primary formation models above). The numbers of nucleated dwarfs that must be accreted by giant ellipticals is also rather intimidating. Moreover, since most dwarf galaxies appear to have their own small globular cluster systems, each accreted nucleus would be accompanied by several globular clusters. Thus even if such a process does add to the globular cluster population of giant ellipticals, most of the additional globulars will not be the nuclei of the dwarf galaxies. Other objections to this picture are given by van den Bergh (1990b).

In Chapter 3, we mentioned that the Sagitarrius dwarf galaxy appears to be currently merging with the Milky Way, bringing with it a handful of globular clusters. While the addition of globular clusters to the Milky Way in this manner does not constitute globular cluster formation, it is possible that many normal galaxies contain some globulars that formed in smaller systems before being accreted by their current host galaxies.

7.5   The importance of environment

Although globular clusters seem to form around all galaxies, the variation in specific frequency between galaxies and its dependence on galaxy properties such as morphology, luminosity, and possibly environment, suggests that some conditions are more conducive to globular cluster formation than others. In Chapter 5 we discussed the high-S_N phenomenon around some giant ellipticals, as well as the higher specific frequencies of globular cluster systems around ellipticals relative to spirals. The interpretation of these results is controversial, but they must nevertheless be explained by any complete model of globular cluster formation. One suggestion by West *et al.* (1995) is that the high-S_N phenomenon can be explained by intergalactic globular clusters. These authors suggest such clusters may either be stripped from galaxies, or that some globular clusters form in the galaxy cluster potential, but are not associated with any individual galaxy.

In the merger picture, the higher specific frequencies around ellipticals are attributed to multiple episodes of globular cluster formation triggered by mergers of gas-rich galaxies. If the merger model is to explain the high-S_N phenomenon, globular cluster formation must be more efficient in the high-S_N galaxies. As mentioned above, destruction mechanisms may also play some role in producing variations in S_N with L (Murali and Weinberg 1996). The possible correlation between specific frequency and the local galaxy density also requires an increase in globular cluster formation efficiency in denser environments.

Part of the difficulty in interpreting evidence from old globular cluster systems is that it is far from obvious what the environmental conditions were like at the time the systems formed. Further, while it is clear that elliptical galaxies have higher specific frequencies than spirals, and that ellipticals are found in regions of higher galaxy density, it is not clear whether the high galaxy density itself is responsible for more efficient globular cluster formation. The balance of the evidence discussed in Chapter 5 suggests that galaxy luminosity may be a more important factor in determining S_N for elliptical galaxies and thus the globular cluster formation efficiency. Since the brighter galaxies that have been studied tend to be found in denser environments, it is difficult to disentangle the two effects.

A different approach to this problem is to determine those environments and associated conditions that are currently producing globular clusters. As stressed earlier, the difficulty with such an approach is the uncertainty concerning young globular cluster candidates and whether they are genuine analogs of their much older counterparts. With this caveat in mind, it is useful to summarize the environments where globular cluster formation *may* be currently occuring.

As discussed in Chapters 4 and 5, there are a number of galaxies in which massive, compact, young star clusters are observed, for which there are no counterparts in the Galaxy. The observations range from a few such young clusters in actively star forming dwarf galaxies, to hundreds or more in ongoing mergers of gas-rich disk galaxies. The common thread uniting the galaxies which host these possible young analogs to globular clusters is that they are *starburst* systems, with locally high rates of star formation and, when measured, high gas densities. Irrespective of the open question of whether some or all of these sites are producing *globular* clusters, we emphasize that there is little doubt that star clusters significantly more massive than those in normal disk galaxies are forming in these environments. Thus we are witnessing conditions that are closer (if not identical) to those responsible for globular cluster formation than are currently prevalent in the Milky Way.

The possibility that galaxy mergers lead to globular cluster formation has received particular attention, primarily because of its implications for theories of galaxy formation and evolution (Schweizer 1987; Ashman and Zepf 1992; also Chapter 6 above). While much of the work in this area has concentrated on the galaxy formation aspects, studies have also been undertaken to address the question of *why* galaxy mergers might lead to globular cluster formation. Kumai *et al.* (1993a) investigated this problem and proposed that collisions of gas clouds within galaxies were required to produce globular clusters (see also Jog and Solomon 1992). The shock and associated compression produced by such collisions is regarded as the salient feature that produces globular clusters. To this extent, the model has similarities with the ideas of Gunn (1980) and McCrea (1982) discussed above. The random velocity of the protoglobular clouds is required to be at least 50–100 km/s in this picture. Further, Kumai *et al.* (1993a) suggest that interactions and mergers between galaxies can produce this kind of motion in gas clouds within the interacting galaxies. In a subsequent paper, Kumai *et al.* (1993b) presented evidence for an increase in specific frequency with the velocity dispersion of the host galaxy, which is related to the increase of S_N with luminosity discussed in Section 5.4 (see also Zepf *et al.* 1994; Djorgovski and Santiago 1992). This is consistent with their picture of globular cluster formation, since the higher-velocity collisions in deeper potential wells are expected to lead to more efficient globular cluster formation. Kumai *et al.* (1993b) also suggest that galaxies in denser environments are more likely to undergo interactions which can increase the random motions of gas clouds within such galaxies. This leads to a predicted increase in S_N with environment. As noted in Chapter 5, the evidence for such a trend is controversial.

While large random motions of gas clouds can be produced by galaxy mergers and interactions, similar conditions are plausible in the protogalactic stage of spiral galaxies. In the context of the merger scenario, this provides a useful link between 'first generation' globular cluster formation in spirals and subsequent merger-driven globular cluster formation in protoellipticals (Ashman and Zepf 1992). However, some of the detailed predictions of the Kumai *et al.* (1993a) picture conflict with recent observations. Like many of the secondary models discussed in Section 7.3 above, the scenario invokes a Jeans condition to imprint a characteristic globular cluster mass. This leads to the prediction of a decrease in globular cluster mass with increasing metallicity, contrary to observation. The Kumai *et al.* (1993a) model also predicts an increase in globular cluster mass with galaxy mass, through the higher-velocity collisions expected in the deeper potential wells of massive galaxies. The invariance of the peak of the globular cluster mass function from one galaxy to another strongly argues against such a relation (see Section 5.3). It is also worth noting that the virial velocity of dwarf galaxies like Fornax is well below that required for the cloud collisions in the Kumai *et al.* (1993a) picture, and yet Fornax has a globular cluster system.

Lee *et al.* (1995) have investigated globular cluster formation in the context of the Mathews and Schramm (1993) galaxy merger model. They show that the properties of the Milky Way halo globulars, such as spatial and metallicity distributions, can be reproduced by collisions of gas clouds at the protogalactic epoch. The details of the globular cluster formation process itself are based primarily on the work of Murray and Lin (1993 and earlier papers) discussed above.

If the young globular cluster candidates observed in several environments are genuine globulars, it seems likely that no single environmental influence is responsible for globular cluster formation. While the jury is still out, it appears to us that globular clusters may form whenever sufficiently large gas clouds form, or when the star formation efficiency in GMCs is enhanced. These conditions are typical of starbursts. Galaxy mergers have long been identified as a prime mechanism for triggering starbursts, but they are not the only one. For the same reasons then, galaxy mergers appear to be an important, but not unique, mechanism for producing globular clusters.

7.6   Future directions

The above discussion illustrates that observations of globular clusters and globular cluster systems already provide tight constraints on acceptable models of globular cluster formation. The approximate universality of the globular cluster luminosity function suggests that globular clusters form in a similar manner in a wide variety of environments, and limits the degree to which long-term dynamical processes can alter an initial population of globular clusters. The lack of correlation between globular cluster metallicity and mass poses problems for models incorporating a cooling criterion to determine the globular cluster mass or the mass function.

Various lines of attack will help refine our current ideas on globular cluster formation. The question of self-enrichment versus pre-enrichment may be clarified by studies of elemental abundances. Can all star-to-star abundance variations be attributed to stellar evolutionary processes, or are some variations primordial? Such issues are also linked to the timescale for globular cluster formation.

Perhaps the central question is: Are globular clusters 'special'? Specifically, do we require a unique formation process for globular clusters, or do they simply represent the high-mass (high-density? high binding energy?) extreme of a continuous process that is essentially the same, from OB associations of a few hundred stars to the largest globular clusters with several million stars?

Whatever the solution to these and other problems, we are more likely to obtain answers if we can study the globular cluster formation process at the present epoch. Thus perhaps the most important question at the time of writing concerns the young globular cluster candidate systems. As described in Chapter 6, the current thrust in this area is to establish whether the luminosity function of these systems is evolving towards a characteristic form of old globular cluster systems. Only if this is convincingly demonstrated can we be confident that these young objects are young globular clusters. Without this evidence, we will be forced to revert to studies of old globular cluster systems to uncover the mysteries of their formation. However, if globular clusters are forming at the present epoch, we have a wonderful opportunity to investigate the process in detail.

8

Future prospects

In this final chapter we consider the promising future directions in globular cluster research. This discussion is driven by those areas where we feel that theoretical and observational progress can have the greatest impact, both in solving currently open questions about globular cluster systems themselves and producing significant advances in other astrophysical fields. To this extent, this chapter is a wish list for the future. The most exciting developments in astrophysics are invariably unexpected, but we feel that offering some speculation on the nature of future advances is useful. We organize this chapter along the lines of the earlier text, starting with globular clusters in the Milky Way and progressing to the distant extragalactic globular cluster systems.

8.1   Milky Way globular clusters

In terms of its impact on other areas of astrophysics and cosmology, the age of Milky Way globular clusters is one of the most important topics of current and future research. There are two primary reasons for this interest. First, globular cluster ages provide a lower limit to the age of the universe, so that when combined with other observable cosmological parameters they provide constraints on the geometry and mass density of the universe. Current age estimates already provide important limits on cosmological models, so reducing uncertainties in these estimates is a key goal. Second, the age *distribution* of Milky Way globular clusters provides critical constraints on the formation history and formation models of the Milky Way.

These two areas are likely to be advanced with different strategies and observations. In principle, a single globular cluster age can provide a useful lower limit to the age of the universe. As discussed in Section 2.2, errors in the distance to a given globular cluster usually dominate observational uncertainities in the derived age. Improvements in this area include refining the calibration of the [Fe/H]–M_V(RR) relation (see Sections 2.2 and 3.1) and reducing uncertainities in reddening corrections. Considerably progress on the [Fe/H]–M_V(RR) calibration is currently underway (see Section 3.1). Deep photometry of globular clusters allows accurate absolute age determinations through main-sequence fitting, and more observational programs in this area are highly desirable. This will be complemented by more accurate determinations of subdwarf absolute magnitudes via parallax, which is required to implement the main-sequence fitting technique.

Relative globular cluster ages are rather more robust against some of the observational issues afflicting absolute ages. The above improvements in absolute age determinations will obviously have an impact in this area, but it is also important to obtain relative ages for a *larger sample* of Milky Way globulars. This will allow advances in constraining formation models of the Milky Way. The current observations seem to point to a picture in which globular clusters in the inner regions of the Galactic halo all formed on a relatively short timescale, whereas clusters in the more remote regions of the Galaxy formed over an appreciable period of time. However, this conclusion has been challenged by ongoing work on the origin of the second parameter problem, discussed further below. Increasing the number of globular clusters with reliable ages is important in establishing whether this general picture is correct and determining the relative fraction of clusters in the two populations. The latter aspect is relevant for constraining conditions during the formation of the Milky Way and possibly for understanding the importance of accretion of material such as satellite galaxies experienced by the Galaxy.

The question of relative ages is linked to the second parameter problem, since age differences between globular clusters have long been regarded as a prime candidate for the second parameter (Section 2.1). Recent results have cast increasing doubt on this idea, however, suggesting that while age probably plays some role, it may not be the dominant second parameter. Improvements in relative age determinations discussed above will be important in resolving this issue, as will advances in stellar evolution theory. If age is *not* the dominant second parameter, further studies between correlations of globular cluster properties and their horizontal branch morphologies will be instrumental in establishing which of the other candidates described in Section 2.1 *is* driving horizontal branch differences.

The chemical properties of globular clusters is another area where there are close links between observation and stellar evolution theory. The chemical homogeneity of iron-peak elements in the vast majority of Galactic globular clusters is one of their most notable features which has had a profound influence on globular cluster formation models (Chapter 7). However, understanding the inhomogeneities in certain elements within globular clusters provides an equally difficult theoretical challenge. As discussed in Section 2.3, the central question is whether these star-to-star variations are primordial or a result of stellar evolutionary processes. The former possibility is difficult to understand in the light of the homogeneity of iron-peak elements, so it seems probable that much theoretical effort will be directed at finding processes that can lead to variations in the observed abundances of elements such as Na and Al. Mixing mechanisms have received the most attention, but these are only believed to be effective in post-main-sequence stars. Consequently, it is important to look for similar variations in these elements amongst main-sequence stars in individual globular clusters.

The dynamics and dynamical evolution of globular clusters has been a lively area of research for some time and will probably remain so. Observational advances in this area owe a lot to *HST* and it seems probable that this trend will continue for the next few years. One area of note is the deep photometry of the stellar luminosity function in globular clusters. Variations in the mass function with distance from the center of globular clusters have already been detected and indicate that mass segregation occurs in such objects. Further studies offer the possibility of revealing the degree of mass segregation, allowing more detailed comparisons with theoretical models and improving our understanding of mass loss from globular clusters. However, even with *HST*, obtaining deep luminosity functions for a significant number of clusters is not an easy task.

As discussed in Chapters 3 and 7, one of the main goals of dynamical studies of the Milky Way globular cluster *system* is to estimate the degree of globular cluster destruction through dynamical processes and thus to determine the original globular cluster population. This area of research is relevant not only to the Milky Way, but to globular cluster systems in general. Through such studies, more reliable limits on the formation efficiency of globular clusters can be obtained, as well as a better understanding of the connection between globular clusters and other stellar populations within galaxies.

Current theoretical research on the dynamical destruction of globular clusters suggests that in the Milky Way and other galaxies the current population may be a small surviving fraction of a more populous original system. One curiosity is that there do not appear to be any definitive observational signatures that significant destruction actually took place. There are clear signs that dynamical evolution is important in the Milky Way system, such as the tendency of collapsed core clusters to be found in the inner regions of the Galaxy (Section 3.5), but essentially no direct observational evidence that most of the globular clusters originally associated with the Milky Way have been destroyed. This is all the more peculiar given that there are several observational signatures that might be expected. As discussed in Section 3.5, dynamical destruction in the Milky Way is most effective for globular clusters at small Galactocentric distances. One therefore expects that the degree of dynamical destruction should be a function of Galactocentric distance and that some evidence of this destruction would be revealed by comparing the properties of globular clusters at different radii. While it does appear that low-concentration clusters are absent from the inner regions of the Galaxy as expected, other signatures are not observed. For example, no significant variation in the GCLF peak with distance from the Galactic center is found, at least if the handful of very remote clusters are excluded from the dataset. This is also the case in the limited amount of data for other galaxies.

Future observational programs may shed light on this conundrum. *HST* observations are particularly valuable since it is relatively straightforward to detect globular clusters fainter than the GCLF peak in a large sample of galaxies. This will enable a detailed study of the variations of the GCLF peak with galactocentric radius. At the same time, further numerical modeling will enable more detailed comparisons with observation to investigate the degree and importance of globular cluster destruction in galaxies.

8.2   Extragalactic globular cluster systems

While some of the dynamical issues and future prospects discussed in the previous section also apply to extragalactic GCSs, there are other areas of research which are unique to such systems, simply because they involve investigating correlations between properties of GCSs and their host galaxies.

There are several outstanding questions at present. Perhaps the most basic involves the total number (or specific frequency) of globular clusters and how it varies with galaxy properties. The fact that ellipticals tend to have higher specific frequencies than spirals is relatively well established, although, as discussed in Chapter 5, the typical difference is only a factor of 2 or so if number per unit stellar mass is considered. The limited number of spiral galaxies in the database is somewhat worrying, but ongoing ground-based and *HST* programs will improve the situation. Such observations will also add significantly to the elliptical galaxy database enabling a better determination of the correlation between S_N and parent galaxy luminosity. Currently it appears that S_N increases with host galaxy luminosity, L, implying more efficient globular cluster formation in higher luminosity ellipticals, but the correlation is not well constrained, partly because of the inhomogeneous nature of the current sample. Further, there is considerable scatter at all L in the value of S_N, so it seems likely that other parameters are also important in determining S_N. Expanding the current database through further observations will hopefully clarify the physical properties that influence globular cluster formation efficiency.

The high-S_N phenomenon has received a considerable amount of attention. The increase of S_N with L suggests the high-S_N systems may not be as anomalous as previously believed, instead reflecting the fact that the high-S_N systems tend to be found around bright galaxies. However, it is also clear that not all bright ellipticals have high values of S_N and that the phenomenon does not appear to correlate with any property of the host galaxy. At present it is difficult to decide whether the high-S_N systems represent a distinct class, or whether they are produced by increased scatter in the S_N–L relation at high luminosities. Given the fact that normal ellipticals with S_N values of 15 or more are only found in the centers of groups and clusters, the former possibility seems more likely, but the data are not conclusive. Resolving this issue requires observations of centrally located ellipticals at great distances to expand the current database. Such observations are planned with *HST*.

The discovery of bimodal and multimodal color distributions in the globular cluster systems of ellipticals represents a major breakthrough in our understanding of these systems. The result has significant implications for the formation of globular clusters and elliptical galaxies which we discuss further in the following two sections. Here we focus on future observational strategies involving studies of extragalactic globular cluster systems that we anticipate will also yield important information.

Establishing the frequency of bimodality in globular cluster color distributions is one important goal that can be achieved by observing more systems. A key element here is to extend studies to lower luminosity ellipticals and those outside clusters, since these objects are underrepresented in the current database. A specific issue of interest is whether faint ellipticals, which often exhibit disky isophotes and surface brightness profiles with steep power-law cores, have globular cluster systems with bimodal color distributions like those seen in brighter ellipticals. This tackles the question of whether there are two types of elliptical galaxy and whether they form in different ways. For example, if the faint ellipticals form primarily from purely dissipative processes, one would expect unimodal rather than multimodal globular cluster color distributions.

A second issue is whether the color gradients in globular cluster systems around ellipticals are purely the result of the red and blue clusters having different spatial profiles, or whether there are metallicity gradients within one or both populations.

The latter result would be evidence for dissipative collapse and enrichment at some point in the galaxy formation process. Further, the detailed nature of the color gradients will allow more specific tests of galaxy formation models.

One relatively unexplored observational area that may provide new insights into these issues is the study of the kinematics of globular cluster systems around ellipticals. Specifically, the combination of velocity measurements with color information will make it possible to search for kinematic differences between globular cluster populations identified by color. In the merger model, the different origins of metal-rich and metal-poor clusters around ellipticals suggest that the two populations may have different kinematic signatures. Computer simulations of mergers that follow the two globular cluster populations are currently underway which may predict observable signatures.

Observations to date have failed to reveal any significant correlations between the mean metallicity and metallicity distribution of globular cluster systems with properties of the host galaxy (see Chapter 5). This result is problematic for all current models of the formation of globular clusters around ellipticals. The ongoing increase in both ground-based and *HST* data concerning the color of globular cluster systems will allow more detailed searches for trends between globular cluster metallicity and galaxy properties, but the lack of obvious trends in current data suggests that detecting any such trends will not be easy. It therefore seems that theoretical modeling of the formation and enrichment processes relevant to globular cluster systems, as well as a more sophisticated inclusion of additional processes such as accretion, may be the most important step forward in this area.

The apparent universality of the GCLF (or the underlying globular cluster mass function) is of considerable interest, both because of the implications for the uniformity of globular cluster formation in different environments and because the peak of the GCLF can be used as a cosmological distance indicator. The primary shortcoming of current observations is that many GCLFs are only observed at magnitudes brighter than the expected peak, so that the magnitude of the peak is inferred rather than observed. This situation is also changing thanks to current observing programs, so that in the next few years it is likely that the universality of the GCLF will be established or variations accurately constrained. Such programs will also allow studies of possible trends between host galaxy properties and properties of the GCLF such as its dispersion, shape, and peak magnitude. Variations of the GCLF, both between galaxies and as a function of position within individual galaxies, are important in studies of the dynamical destruction and evolution of globular cluster systems, and for distance scale measurements.

8.3   Galaxy formation

In Chapter 6 we discussed the active interplay between observations of extragalactic globular cluster systems and the constraints they provide on galaxy formation models. We anticipate that this connection will persist in the future. In the case of elliptical galaxies, a central question is the importance that galaxy mergers played in their formation.

In testing any model, a central goal is to find an observable that provides a definitive test between competing models. Initially it appeared that the color distribution of globular cluster systems provided such an observable, since only the merger model predicted bimodal and multimodal color distributions. The detection of such color distributions seems to rule out traditional single-collapse scenarios. However, there have been some suggestions as to how more elaborate non-merger models might be constructed which may explain these color distributions. Further, it has been realized that the merger model requires added features to explain some of the details of globular cluster systems around ellipticals and elliptical galaxies themselves.

If mergers do produce elliptical galaxies, the progenitors must have been considerably more gas-rich than current day spirals. This is primarily because the stars in ellipticals have higher [O/Fe] ratios than those in spirals, so that the putative mergers are required to occur before most of the gas in the progenitor spirals has formed stars. This does not alter the prediction of bimodal color distributions for the globular clusters of ellipticals formed in this manner, but it is worth keeping in mind that the merger model requires progenitors that are somewhat different to galaxies at the present epoch. Such differences are actually expected both on theoretical and observational grounds.

A more critical issue is the expectation from cosmological models that protogalaxies are lumpy; that is, they are comprised of many lower mass, possibly independent, overdense regions. While this lumpiness is apparent in numerical simulations of galaxy formation and has been included in some analytical work on galaxy formation, it has not been widely incorporated into studies of globular cluster formation in such protogalaxies. This is the main factor that may blur (but not remove) the distinction between the merger and single-collapse models. This is because in a lumpy protogalaxy there will be collisions between individual lumps so that some level of 'merging' will occur in a single protogalaxy. Further, the young spirals of the merger model may retain some of this initial lumpiness, in contrast to the well-ordered gas disks of present-day spirals. There is still a clear distinction here between models: the merger model of elliptical galaxy formation includes a major merger of two, roughly equal mass gas-rich galaxies, whereas no such major mergers are invoked in the single-collapse picture. The question is less clear cut when one considers the possibility of several gas-rich protogalaxies coming together at the same time to form a galaxy. This is neither a classic disk-plus-disk merger, nor a single collapse, but the situation may occur quite frequently in the universe.

While the distinction between galaxy formation models is diminishing, it is interesting to note that this is partly due to the accumulating evidence obtained from globular cluster systems. Further, we feel that globular clusters still provide a discriminant between models. Bimodal and multimodal color distributions indicate two bursts of star formation *organized over the bulk of the galaxy*. Such distributions also imply that whatever type of objects were involved in building the galaxy had already formed globular clusters *before* they became a part of the elliptical galaxy observed today. In a single, lumpy protogalaxy in which clouds merge to form stars and globular clusters, it seems to us difficult to produce distinct peaks in the globular cluster color distribution. Even if different 'mini-mergers' did produce different levels of enrichment, the process would be local and would not give rise to global color bimodality. The challenge for such single-collapse models is to reproduce bimodal color distributions in a physically well-motivated way. That is not to say that the merger picture does not have problems: as we discussed in Chaper 6 it currently suffers from shortcomings. However, in our view, the globular cluster color distributions of many ellipticals indicate, in a model-independent way, that two bursts of star formation occurred, and currently the only galaxy formation model that naturally predicts such a situation for elliptical galaxies is the merger picture.

The above considerations suggest that the recent increase in observational information about globular cluster systems has outstripped theoretical work in this area. Thus while further observations of globular cluster systems will be important in understanding the formation of elliptical galaxies, it is also apparent that advances are also required on the theoretical front. One important area mentioned above is the use of computer simulations of mergers in which globular clusters from the progenitor spirals and those formed in the merger are tracked. This will allow more detailed predictions of the kinematic and spatial distributions of globular cluster systems around ellipticals in the context of the merger model. More detailed work on the enrichment history of globular cluster systems is also desirable, within both the merger and non-merger models.

8.4   Globular cluster formation

It is somewhat ironic, given our claims that globular clusters are valuable tools in solving many astrophysical questions, that the formation of these systems is still poorly understood. As discussed in Chapters 5 and 7, a critical element in furthering this area of study is to establish whether the young globular cluster candidates found in galaxy mergers and other environments are genuine globulars. It seems clear that some of these candidates have exactly the properties expected of young globular clusters. However, one argument *against* the identification of young massive star clusters with young globulars is based on the cluster luminosity function. As discussed in Chapter 5, many of the low luminosity clusters found in young cluster systems must dissolve in some way over the next Hubble time if the luminosity function of these systems is to resemble the log-normal luminosity function of old globular clusters. There are various dynamical processes that could, in principle, cause such evolution of the luminosity function, but, as discussed in Chapters 5 and 7, there is little evidence that such processes have had a significant effect on modifying the GCLF.

Given current evidence, we suspect that a more promising possibility for producing luminosity function evolution is that stellar evolutionary processes disrupt clusters while they are still relatively young. This possibility may be tested empirically by studying the luminosity function of young cluster systems with a range of ages. If the luminosity function is found to turn over in the older systems (ages around 10^9 years), it will be a strong indication that stellar evolutionary processes are responsible for cluster disruption. More importantly, it will provide compelling evidence that at least some of the massive, young star clusters observed in active star-forming regions are young globular clusters. Moreover, if these young clusters are confirmed to be young globulars, we will have the opportunity to study the process of globular cluster formation directly. This will give formation models a firm empirical basis. Further, such a finding will imply that the formation of globular clusters is more closely related to the formation of other star clusters than has often been supposed in the past. While constructing a complete formation model of any type of star cluster may still prove a highly complex and challenging task, it will surely be made easier if the formation process can be observed.

Appendix

This appendix contains data on extragalactic globular clusters systems and their host galaxies. These data are used extensively in Chapters 5 and 6 for studying trends and correlations in these properties. Similar information can be found in Harris and Harris (1997) and Kissler-Patig (1997).

An asterisk after the galaxy number indicates that less than 10% of the estimated number of globular clusters in that galaxy have been observed. All galaxy numbers are NGC numbers unless otherwise specified.

The sources listed in the right-hand column can be found in the references of the main text through the following key:

Ba95, Baum *et al.* (1995); Bl95, Blakeslee and Tonry (1995); Bo92, Bothun *et al.* (1992); Br92, Bridges and Hanes (1992); Br94, Bridges and Hanes (1994); Br96, Bridges *et al.* (1996); Br97, Bridges *et al.* (1997) Bu92, Butterworth and Harris (1992); Ch3, chapter 3 of this book; Ch4, chapter 4 of this book Ch93, Christian (1993) Di96, Dirsch (1996); Du96a, Durrell *et al.* (1996a), Du96b, Durrell *et al.* (1996b); Fl95, Fleming *et al.* (1995), Fo96; Forbes *et al.* (1996); Ge96, Geisler *et al.* (1996); H91, Harris (1991); Ha92, Harris *et al.* (1992); Ha95, Harris *et al.* (1995); Ho95, Hopp *et al.* (1995); Hi96, Hilker and Kissler-Patig (1996); Ka96, Kaisler *et al.* (1996); Ki94, Kissler *et al.* (1994); KP96, Kissler-Patig *et al.* (1996); KP97, Kissler-Patig *et al.* (1997); Le94, Lee *et al.* (1994) McL94, McLaughlin *et al.* (1994); McL95, McLaughlin *et al.* (1995); Os93, Ostrov *et al.* (1993); Pe95, Perelmutter and Racine (1995); Ri92, Richtler *et al.* (1992); Se95, Secker *et al.* (1995); Sh93, Schommer (1993); Sc93, Suntzeff (1993); W97, Whitmore *et al.* (1997) Ze94, Zepf *et al.* (1994), Ze95, Zepf *et al.* (1995)

Some properties of globular cluster systems and their host galaxies

Galaxy	Type	$(m-M)_V$	M_V	N_{GC}[a]	S_N	α_{proj}	$[Fe/H]_{GCS}$	Source
Galaxy	Sbc		−21.3 :	180 ± 20	0.5 ± 0.1	−2, −2.5	−1.34 ± 0.07	Ch3
LMC	Sm	18.7	−18.4	13:	0.6 ± 0.2		−1.8 ± 0.2	Ch4, Sz93
SMC	Im	18.9	−16.9	1:	0.2:		−1.4	Ch4
Fornax	dE0	20.6	−13.7	5:	17 ± 7		−1.9 ± 0.2	Ch4, H91
147	dE5	24.6	−15.1	4:	3.6 ± 2		−2.05 ± 0.4	H91
185	dE3	24.6	−15.4	8:	5.5 ± 2		−1.65 ± 0.25	H91
205	dE5	24.6	−16.5	9:	2.3 ± 1		−1.45 ± 0.1	H91
221	E2	24.6	−16.5	0:	< 1			H91
M31	Sb	24.6	−21.8	450 ± 100	0.9 ± 0.2	−2.5	−1.21 ± 0.07	Ch4
M33	Scd	24.8	−19.4	30:	0.5 ± 0.2		−1.6 ± 0.2	Ch93, Sh93
55	Sm	25.7:	−18.6	2:				H91
253	Sc	26.7:	−20.2	24:	0.2 ± 0.1			H91
524	S0	32.7	−22.5	3300 ± 1000	3.3 ± 1.0	−1.7 ± 0.1		H91
720	E5	31.4	−21.2	660 ± 190	2.2 ± 0.6	−2.2 ± 0.1		KP96
1052	E4	31.3	−20.9	430 ± 80	1.9 ± 0.4	−2.3 ± 0.3		H91
1275*	Ep	34.9	−23.0	8000 ± 3000	5 ± 2			Ka96
1374	E0	31.1	−20.0	410 ± 82	4.1 ± 0.8	−1.8 ± 0.3	−0.54 ± 0.30	KP97
1379	E0	31.1	−20.1	314 ± 63	2.9 ± 0.6	−2.1 ± 0.6	−0.25 ± 0.30	KP97
1387	S0	31.1	−20.4	389 ± 110	2.7 ± 0.8	−2.2 ± 0.3	−0.25 ± 0.40	KP97
1399	E1/cD	31.1	−21.6	5410 ± 900	12.4 ± 3	−1.6 ± 0.2	−0.75 ± 0.15	KP97, Os93
1404	E1	31.1	−21.1	880 ± 140	3.2 ± 0.5	−2.0 ± 0.1		R92
1427	E5	31.1	−20.2	510 ± 87	4.2 ± 0.7	−2.0 ± 0.3	−0.94 ± 0.25	W97, Fo96, KP97
1439	E1	31.3	−20.0				−1.05 ± 0.25	W97, Fo96
1549	E0	30.7	−20.9	140 ± 60	0.6 ± 0.3	−1.8 ± 0.3		H91
1553	S0	30.7	−21.3	450 ± 100	1.4 ± 0.3	−2.3 ± 0.3		H91
1700	E4	33.6	−22.5				−1.05 ± 0.40	W97, Fo96
3031	Sab	27.8	−21.1	210 ± 30	0.7 ± 0.1	−2.1	−1.48 ± 0.2	Pe95
2683	Sb	29.8	−20.8	310 ± 100	1.7 ± 0.5			H91
3109	Im	25.6	−16.5	20::	5 : ±2			H91
3115	S0	30.2	−21.3	520 ± 120	1.6 ± 0.4	−1.8 ± 0.2		H91
3115B	dE1,N	30.2	−17.7	59 ± 23	4.9 ± 1.9	−1.8 ± 0.4	−1.2 ± 0.13	Du96a
3226	E2	31.0:	−19.4	470 ± 200	8.2 ± 3.5	−2.5 ± 0.5		H91
3311	E0/cD	33.6	−22.3	12500 ± 5000	15 ± 6	−1.3 ± 0.2	−0.31 ± 0.15	Se95, McL95
3377	E5	30.3	−20.0	240 ± 60	2.4 ± 0.6	−1.9 ± 0.2		H91
3379	E1	30.3	−21.0	300 ± 160	1.2 ± 0.6	−1.8 ± 0.2		H91
3384	S0	30.3	−20.4	125 ± 70	0.9 ± 0.5			H91
3557*	E3	33.0	−22.8	400 ± 300	0.3 ± 0.2			H91
3607*	S0	30.8	−20.9	800 ± 600	3.5 ± 2.6	−2.6 ± 0.5		H91
3608	E2	30.9	−20.1				−1.05 ± 0.25	W97, Fo96
3842*	E3	34.8	−22.8	10500 ± 4000	8 ± 3	−1.2 ± 0.2		Bu92
3923*	E3	31.9	−22.0	4300 ± 1000	6.8 ± 1.6	−1.4 ± 0.2	−0.56 ± 0.15	Z94, Z95
4073*	E1/cD	34.6	−23.1	7000 ± 2000	4 ± 1.2	−1.0 ± 0.3		Br94
4278	E1	30.6	−20.4	1000 ± 300	6.9 ± 2.1	−1.9 ± 0.2	−0.79 ± 0.25	Fl95, W97, Fo96
4565	Sb	30.1	−21.5	180 ± 45	0.5 ± 0.1	−2.3 ± 0.6		Fl95, H91
4216	Sb	31.0	−21.8	620 ± 310	1.2 ± 0.6			H91
4340*	SB0	31.0	−19.9	775 ± 310	8.5 ± 3.1			H91
4365	E2	31.4	−21.9	2500 ± 200	4.3 ± 0.6	−1.2 ± 0.3	−0.79 ± 0.25	H91, W97, Fo96
4374*	E1	31.0	−21.9	3040 ± 800	5.3 ± 1.4			H91
4406*	E3	31.0	−22.2	3350 ± 900	4.4 ± 1.2		−1.20 ± 0.25	H91, W97, Fo96
4472	E2	31.0	−22.6	6300 ± 1900	5.6 ± 1.7	−1.3 ± 0.2	−0.93 ± 0.15	Ge96
4486	E0	31.0	−22.4	13000 ± 500	14 ± 3	−1.5 ± 0.2	−0.94 ± 0.15	Le94, McL94
4494	E1	30.5	−20.8	1000 ± 350	5.2 ± 1.4	−1.1 ± 0.4	−0.99 ± 0.25	W97, Fo96
4526*	S0	31.0	−21.4	2700 ± 800	7.4 ± 2.2			H91
4564*	E6	31.0	−19.9	1000 ± 300	11 ± 3.4			H91
4569*	Sab	31.0	−21.7	900 ± 300	1.9 ± 0.6			H91
4589	E2	31.7	−21.0				−0.74 ± 0.25	W97, Fo96
4596*	SB0	31.0	−20.4	2800::	13::			H91
4621*	E5	31.0	−21.5	1900 ± 400	4.8 ± 1.2			H91
4636	E0	31.0	−21.5	3000 ± 500	7.5 ± 1.0	−1.0 ± 0.1		Ki94
4649*	E2	31.0	−22.2	5100 ± 1100	6.7 ± 1.4			H91
4697*	E6	30.8	−21.6	1100 ± 400	2.5 ± 1.0	−1.9 ± 0.2		Di96
VCC1254	dE0,N	31.0	−16.4	28 ± 13	7.7 ± 3.6		−1.45 ± 0.24	Du96b

Some properties of globular cluster systems and their host galaxies (continued)

Galaxy	Type	$(m-M)_V$	M_V	N_{GC}	S_N	α_{proj}	$[Fe/H]_{GCS}$	Source
VCC1386	dE3,N	31.0	−16.9	26 ± 12	4.5 ± 2.1		−1.45 ± 0.22	Du96b
4594	Sa	29.8	−22.2	1600 ± 800	2 ± 1	−1.8 ± 0.2	−0.8 ± 0.25	H91, Br92, Br97
5018	E4p	33.4	−22.8	1200 ± 500	0.9 ± 0.4	−1.3 ± 0.4		Hi96
5170	Sb	31.7:	−21.6	400 ± 150	0.9 ± 0.3	−1.7 ± 0.3		H91
4874	E0	34.9	−23.0	22600 ± 5200	14.3 ± 3.3			Bl95
4881	E0	34.9	−21.6	400 ± 80	1.0 ± 0.2			Ba95
4889*	E4	34.9	−23.5	17300 ± 4500	6.9 ± 1.8			Bl95
5128	E0	28.3	−22.0	1700 ± 400	2.6 ± 0.6		−0.98 ± 0.15	H91, Ha92
5322	E3	32.1	−21.9				−0.74 ± 0.25	W97, Fo96
5629	cD	34.0	−21.8	< 2500	< 5			Br94
5813	E1	32.0	−21.5	2400 ± 600	6.0 ± 1.5	−2.2 ± 0.3	−0.99 ± 0.25	Ho95, W97, Fo96
5846*	E0	32.0	−22.0	2200 ± 1300	3.5 ± 2.1			H91
5982	E3	32.9	−21.7				−0.99 ± 0.25	W97, Fo96
UGC9799	E/cD	35.9	−23.4	48000 ± 16000	21 ± 7	−1.4 ± 0.3		Ha95
UGC9958	E/cD	36.3	−23.4	27000 ± 13000	12 ± 6	−1.1 ± 0.5		Ha95
6166	E2/cD	35.6	−23.6	30000 ± 14000	11 ± 5	−1.0 ± 0.1	−1.0 ± 0.4	Br96
7626	E1	33.4	−22.3				−0.74 ± 0.25	W97, Fo96
7768	E2/cD	35.3	−22.9	4050 ± 2600	2.8 ± 1.8	−1.3 ± 0.3		Ha95
7814	Sab	30.6	−20.4	500 ± 160	3.5 ± 1.1			Bo92
IC1459	E3	31.7	−21.8				−0.69 ± 0.25	W97, Fo96

[a] Colons indicate that the globular cluster surveys may be incomplete.

References

Abraham, R.G. and van den Bergh, S., 1995, *Astrophys. J.*, **438**, 218.

Aguilar, L., Hut, P., and Ostriker, J.P., 1988, *Astrophys. J.*, **335**, 720.

Ajhar, E.A., Blakeslee, J.P., and Tonry, J.L., 1994, *Astron. J.*, **108**, 2087.

Ajhar, E.A., Grillmair, C.J., Lauer, T.R., Baum, W.A., Faber, S.M., Holtzman, J.A., Lynds, C.R., and O'Neil Jr, E.J., 1996, *Astron. J.*, **111**, 1110.

Alexander, D.R., Brocato, E., Cassisi, S., Castellani, V., Ciacio, F., and Degl'Innocenti, S., 1996, *Astron. Astrophys.*, in press.

Allen, C., Schuster, W.J., and Poveda, A., 1991, *Astron. Astrophys.*, **244**, 280.

Anthony-Twarog, B.J., Twarog, B.A., and Craig, J., 1995, *Publ. Astron. Soc. Pac.*, **107**, 32.

Arimoto, N. and Yoshii, Y., 1987, *Astron. Astrophys.*, **173**, 23.

Armandroff, T.E., 1989, *Astron. J.*, **97**, 375.

Armandroff, T.E., 1993, in *The Globular Cluster–Galaxy Connection*, eds. G.H. Smith and J.P. Brodie (ASP, San Francisco), p. 48.

Armandroff, T.E., Da Costa, G.S., and Zinn, R., 1992, *Astron. J.*, **104**, 164.

Armandroff, T.E. and Zinn, R., 1988, *Astron. J.*, **96**, 92.

Arp, H.C., 1955, *Astron. J.*, **60**, 317.

Arp, H.C., Baum, W.A., and Sandage, A., 1953, *Astron. J.*, **58**, 4.

Arp, H.C. and Sandage, A., 1985, *Astron. J.*, **90**, 1163.

Ashman, K.M., 1990, *Mon. Not. R. astr. Soc.*, **247**, 662.

Ashman, K.M., 1992, *Publ. Astron. Soc. Pac.*, **104**, 1109.

Ashman, K.M. and Bird, C.M., 1993, *Astron. J.*, **106**, 2281.

Ashman, K.M., Bird, C.M., and Zepf, S.E., 1994, *Astron. J.*, **108**, 2348.

Ashman, K.M. and Carr, B.J., 1988, *Mon. Not. R. astr. Soc.*, **234**, 219.

Ashman, K.M. and Carr, B.J., 1991, *Mon. Not. R. astr. Soc.*, **249**, 13.

Ashman, K.M., Conti, A., and Zepf, S.E., 1995, *Astron. J.*, **110**, 1164.

Ashman, K.M. and Zepf, S.E., 1992, *Astrophys. J.*, **384**, 50.

Baade, W., 1958, in *Stellar Populations*, ed. D.J.K. O'Connell (North Holland, Amsterdam), p. 303.

Baade, W. and Hubble, E., 1939, *Publ. Astron. Soc. Pac.*, **51**, 40.

Bahcall, J.N. and Tremaine, S., 1981, *Astrophys. J.*, **244**, 805.

Baikie, Rev. J., 1911, in *Peeps at the Heavens* (Adam and Charles Black, London).

Bailyn, C.D., 1992, *Astrophys. J.*, **392**, 519.

Bailyn, C.D. 1995, *Ann. Rev. Astron. Astrophys.*, **33**, 133.

Bailyn, C.D. and Pinsonneault, M.H., 1995, *Astrophys. J.*, **439**, 705.

Baraffe, I., Charbrier, G., Allard, F., and Hauschildt, P.H., 1995, *Astrophys. J. Lett.*, **446**, L35.

Barnes, J.E. and Hernquist, L., 1992, *Ann. Rev. Astron. Astrophys.*, **30**, 705.

Barth, A.J., Ho, L.C., Filippenko, A.V., and Sargent, W.L.W., 1995, *Astron. J.*, **110**, 1009.

Battistini, P.L., Bònoli, F., Casavecchia, M., Ciotti, L., Federici, L., and Fusi Pecci, F., 1993, *Astron. Astrophys.*, **272**, 77.

Baum, W.A., 1955, *Publ. Astron. Soc. Pac.*, **67**, 328.

Baum, W.A., Hammegren, M., Groth, E.J., Ajhar, E.A., Lauer, T.R., O'Neil, Jr, E.J., *et al.*, 1995, *Astron. J.*, **110**, 2537.

Beauchamp, D., Hardy, E., Suntzeff, N.B., and Zinn, R., 1995, *Astron. J.*, **109**, 1628

Beers, T.C., Flynn, K., and Gebhardt, K., 1990, *Astron. J.*, **100**, 32.

Bell, R.A., Hesser, J.E., and Cannon, R.D., 1983, *Astrophys. J.*, **269**, 580.

Bendinelli, O., Cacciari, C., Djorgovski, S.G., Federici, L., Ferraro, F., Fusi Pecci, F., Parmeggiani, G., Weir, N., and Zavatti, F., 1993, *Astrophys. J. Lett.*, **409**, L17.

Benz, W. and Hills, J.G., 1987, *Astrophys. J.*, **323**, 614.

Bergbusch, P.A. and VandenBerg, D.A., 1992, *Astrophys. J. Suppl.*, **81**, 163

Berman, V.G. and Suchkov, A.A., 1991, *Astrophys. Space Sci.*, **184**, 169.

Bhatia, R.K. and MacGillivary, H.T., 1988, *Astron. Astrophys.*, **211**, 9.

Binney, J.J., 1977, *Astrophys. J.*, **215**, 483.

Binney, J.J., 1981, in *The Structure and Evolution of Normal Galaxies*, eds. S.M. Fall and D. Lynden-Bell (Cambridge University Press), p. 55.

Binney, J.J. and Tremaine, S., 1987, *Galactic Dynamics* (Princeton University Press, Princeton).

Bird, C.M. and Beers, T.C., 1993, *Astron. J.*, **105**, 1596.

Blakeslee, J.P. and Tonry, J.L. 1995, *Astrophys. J.*, **442**, 579.

Blakeslee, J.P. and Tonry, J.L. 1996, *Astrophys. J. Lett.*, **465**, L19.

Blumenthal, G.R., Faber, S.M., Primack, J.R., and Rees, M.J., 1984, *Nature*, **311**, 517.

Bolte, M., 1989, *Astron. J.*, **97**, 1688.

Bolte, M., 1992, *Astrophys. J. Suppl.*, **82**, 145.

Bolte, M., 1993, in *The Globular Cluster–Galaxy Connection*, eds. G.H. Smith and J.P. Brodie (ASP, San Francisco), p. 60.

Bolte, M. and Hogan, C.J., 1995, *Nature*, **376**, 399.

Bothun, G.D., Harris, H.C., and Hesser, J.E., 1992, *Publ. Astron. Soc. Pac.*, **104**, 1220.

Bridges, T.J., Hanes, D.A., and Harris, W.E., 1991, *Astron. J.*, **101**, 469.

Bridges, T.J. and Hanes, D.A., 1992, *Astron. J.*, **103**, 800.

Bridges, T.J. and Hanes, D.A., 1994, *Astrophys. J.*, **431**, 625.

Bridges, T.J., Carter, D., Harris, W.E., and Pritchet, C.J., 1996a, *Mon. Not. R. astr. Soc.*, **281**, 1290.

Bridges, T.J., Ashman, K.M., Zepf, S.E., Carter, D., and Hanes, D.A., 1997, *Mon. Not. R. astr. Soc.*, **284**, 376.

Briley, M.M., Hesser, J.E., and Bell, R.A., 1991, *Astrophys. J.*, **373**, 482.

Brodie, J.P., 1993, in *The Globular Cluster–Galaxy Connection*, eds. G.H. Smith and J.P. Brodie (ASP, San Francisco), p. 483.

Brodie, J.P. and Huchra, J.P., 1990, *Astrophys. J.*, **362**, 503.

Brodie, J.P. and Huchra, J.P., 1991, *Astrophys. J.*, **379**, 157.

Brosche, P. and Lentes, F.-T., 1984, *Astron. Astrophys.*, **139**, 474.

Brown, J.H., Burkert, A., and Truran, J.W., 1991, *Astrophys. J.*, **376**, 115.

Bruzual, A.G. and Charlot, S., 1996, private communication.

Buonanno, R., 1993, in *The Globular Cluster–Galaxy Connection*, eds. G.H. Smith and J.P. Brodie (ASP, San Francisco), p. 131.

Buonanno, R., Buscema, G., Fusi Pecci, F., Richer, H.B., and Fahlman, G.G., 1990, *Astron. J.*, **100**, 1811.

Buonanno, R., Buscema, G., Fusi Pecci, F., Richer, H.B., and Fahlman, G.G., 1991, in *The Formation and Evolution of Star Clusters*, ed. K. Janes (ASP, San Francisco), p. 240.

Buonanno, R., Corsi, C.E., and Fusi Pecci, F., 1985, *Astron. Astrophys.*, **145**, 97.

Burkert, A., Truran, J.W., and Hensler, G., 1992, *Astrophys. J.*, **391**, 651.

Burstein, D., Faber, S.M., Gaskell, C.M., and Krumm, N., 1984, *Astrophys. J.*, **287**, 586.

Butterworth, S.T. and Harris, W.E. 1992, *Astron. J.*, **103**, 1828.

Cacciari, C., Cassatella, A., Bianchi, L., Fusi Pecci, F., and Kron, R.G., 1982, *Astrophys. J.*, **261**, 77.

Capaccioli, M., Ortolani, S., and Piotto, G., 1991, *Astron. Astrophys.*, **244**, 298.

Capaccioli, M., Piotto, G., and Stiavelli, M., 1993, *Mon. Not. R. astr. Soc.*, **261**, 819.

Caputo, F., and Castellani, V., 1984, *Mon. Not. R. astr. Soc.*, **207**, 185.

Carlberg, R.G., 1984a, *Astrophys. J.*, **286**, 403.

Carlberg, R.G., 1984b, *Astrophys. J.*, **286**, 416.

Carlberg, R.G. and Pudritz, R.E., 1990, *Mon. Not. R. astr. Soc.*, **247**, 353.

Carney, B.W., 1996, in *Formation of the Galactic Halo...Inside and Out*, eds. H. Morrison and A. Sarajedini (ASP, San Francisco), p. 103.

Carney, B.W., Laird, J.B., Latham, D.W., and Aguilar, L.A., 1996, *Astron. J.*, **112**, 668.

Carr, B.J. and Rees, M.J., 1984, *Mon. Not. R. astr. Soc.*, **206**, 315.

Cavaliere, A., and Padovani, P., 1989, *Astrophys. J. Lett.*, **340**, L5.

Cayrel, R., 1986, *Astron. Astrophys.*, **168**, 81.

Chaboyer, B., 1995, *Astrophys. J. Lett.*, **444**, L9.

Chaboyer, B., Demarque, P., and Sarajedini, A., 1996, *Astrophys. J.*, **459**, 558.

Chaboyer, B., Sarajedini, A., and Demarque, P., 1992, *Astrophys. J.*, **394**, 515.

Charlot, S. and Bruzual, A.G., 1996, private communication.

Charlot, S., Worthey, G., and Bressan, A., 1996, *Astrophys. J.*, **457**, 625.

Chernoff, D. and Djorgovski, S., 1989, *Astrophys. J.*, **339**, 904.

Chernoff, D., Kochanek, C.S., and Shapiro, S.L., 1986, *Astrophys. J.*, **309**, 183.

Chernoff, D. and Shapiro, S.L., 1987, *Astrophys. J.*, **332**, 113.

Chernoff, D. and Weinberg, M.D., 1990, *Astrophys. J.*, **351**, 121.

Christian, C.A., 1993, in *The Globular Cluster–Galaxy Connection*, eds. G.H. Smith and J.P. Brodie (ASP, San Francisco), p. 448.

Christian, C.A. and Schommer, R.A., 1988, *Astron. J.*, **95**, 704.

Cohen, J.G., 1976, *Astrophys. J. Lett.*, **203**, L127.

Cohen, J.G., 1978, *Astrophys. J.*, **223**, 487.

Cohen, J.G., 1988, *Astron. J.*, **95**, 682.

Cohen, J.G., 1992, *Astrophys. J.*, **400**, 528.

Cohen, J.G., Persson, S.E., and Searle, L., 1984, *Astrophys. J.*, **281**, 141.

Cohn, H. and Hut, P., 1984, *Astrophys. J. Lett.*, **277**, L45.

Cole, S., Aragin-Salamanca, A., Frenk, C.S., Navarro, J.F., and Zepf, S.E., 1994, *Mon. Not. R. astr. Soc.*, **271**, 781.

Cool, A.M., Piotto, G., and King, I.R. 1996, *Astrophys. J.*, **485**, 655.

Cool, A.M., Grindlay, J.E., Cohn, H.N., Lugger, P.M., and Slavin, S.D., 1995, *Astrophys. J.*, **439**, 695.

Côté, P., Pryor, C., McClure, R.D., Fletcher, J.M., and Hesser, J.E., 1996, *Astron. J.*, 112, 574.

Côté, P., Welch, D.L., Fischer, P., Da Costa, G.S., Tamblyn, P., Seitzer, P., and Irwin, M.J., 1994, *Astrophys. J. Suppl.*, **90**, 83.

Cottrell, P.L. and Da Costa, G.S., 1981, *Astrophys. J. Lett.*, **245**, L79.

Couture, J., Harris, W.E., and Allwright, J.W.B., 1990, *Astrophys. J Suppl.*, **73**, 671.

Couture, J., Harris, W.E., and Allwright, J.W.B., 1991, *Astrophys. J.*, **372**, 97.

Covino, S. and Pasinetti Fracassini, L.E., 1993, *Astron. Astrophys.*, **270**, 83.

Cowley, A.P. and Burstein, D., 1988, *Astron. J.*, **95**, 1071.

Crampton, D., Schade, D.J., Chayer, P., and Cowley, A.P., 1985, *Astrophys. J.*, **228**, 494.

Croswell, K., Latham, D.W., Carney, B.W., Schuster, W., and Aguilar, L., 1991, *Astron. J.*, **101**, 2078.

Crotts, A.P.S., Kron, R.G., Cacciari, C., and Fusi Pecci, F., 1990, *Astron. J.*, **100**, 141.

Cudworth, K.M. and Hanson, R.B., 1994, *Astron. J.*, **105**, 168.

Da Costa, G.S., 1991, in *IAU Symposium No. 148, The Magellanic Clouds*, eds. R.F. Haynes and D.K. Milne (Kluwer, Dordrect), p. 183.

Da Costa, G.S., 1993, in *The Globular Cluster–Galaxy Connection*, eds. G.H. Smith and J.P. Brodie (ASP, San Francisco), p. 363.

Da Costa, G.S., Armandroff, T.E., and Norris, J.E., 1992, *Astron. J.*, **104**, 154.

D'Antona, F. and Mazzitelli, I., 1996, *Astrophys. J.*, **456**, 329.

Davidge, T.J., 1990, *Astrophys. J. Lett.*, **351**, L37.

Davies, R.L., Efstathiou, G., Fall, S.M., Illingworth, G., and Schechter, P.L., 1983, *Astrophys. J.*, **266**, 41.

Davies, R.L., Sadler, E.M., and Peletier, R.F., 1993, *Mon. Not. R. astr. Soc.*, **262**, 650.

Dejonghe, H. and Merritt, D., 1992, *Astrophys. J.*, **391**, 531.

De Marchi, G. and Paresce, F., 1994, *Astrophys. J.*, **422**, 597.

De Marchi, G. and Paresce, F., 1995a, *Astron. Astrophys.*, **304**, 202.

De Marchi, G. and Paresce, F., 1995b, *Astron. Astrophys.*, **304**, 211.

Demarque, P., 1980, in *Star Clusters*, ed. J.E. Hesser (Reidel, Dordrecht), p. 281.

de Young, D.S., Lind, K., and Strom, S.E., 1983, *Publ. Astron. Soc. Pac.*, **95**, 410.

de Zeeuw, T., 1985, *Mon. Not. R. astr. Soc.*, **216**, 273.

Dinescu, D.I., Girard, T.M., van Altena, W.F., and López, C.E. 1996, in *Formation of the Galactic Halo...Inside and Out*, eds. H. Morrison and A. Sarajedini (ASP, San Francisco), p. 261.

Dirsch, B. 1996, Master's Thesis, Univ. of Bonn

Di Stefano, R. and Rappaport, S., 1994, *Astrophys. J.*, **423**, 274.

Djorgovski, S.G., 1993, in *The Globular Cluster–Galaxy Connection*, eds. G.H. Smith and J.P. Brodie (ASP, San Francisco), p. 496.

Djorgovski, S.G., 1995, *Astrophys. J. Lett.*, **438**, L29.

Djorgovski, S.G. and Davis, M., 1987, *Astrophys. J.*, **313**, 59.

Djorgovski, S.G., Gal, R.R., McCarthy, J.K., Cohen, J.G., de Carvalho, R.R., Meylan, G., Bendinelli, O., and Parmeggiani, G., 1997, *Astrophys. J. Lett.*, **474**, L19.

Djorgovski, S.G. and King, I.R., 1986, *Astrophys. J. Lett.*, **305**, L61.

Djorgovski, S.G. and Meylan, G., 1994, *Astron. J.*, **108**, 1292.

Djorgovski, S.G., Piotto, G., and Capaccioli, M., 1993, *Astron. J.*, **105**, 2148.

Djorgovski, S.G. and Santiago, B.X., 1992, *Astrophys. J. Lett.*, **391**, L85.

Dopita, M.A. and Smith, G.H., 1986, *Astrophys. J.*, **304**, 283.

Dressler, A. Lynden-Bell, D., Burstein, D., Davies, R.J., Faber, S.M., Terlevich, R.J., and Wegner, G., 1987, *Astrophys. J.*, **313**, 42.

Dubath, P., Mayor, M., and Meylan, G., 1993, in *The Globular Cluster–Galaxy Connection*, eds. G.H. Smith and J.P. Brodie (ASP, San Francisco), p. 557.

Dubath, P., Meylan, G., and Mayor, M., 1992, *Astrophys. J.*, **400**, 510.

Dubath, P., Meylan, G., Mayor, M., and Magain, P., 1990, *Astron. Astrophys.*, **239**, 142.

Durrell, P.R., Harris, W.E., Geisler, D., and Pudritz, R.E., 1996a, *Astron. J.*, **112**, 972.

Durrell, P.R., McLaughlin, D.E., Harris, W.E., and Hanes, D.A., 1996b, *Astrophys. J.*, **463**, 543.

Edmonds, P.D., Gilliland, R.L., Guhathakurta, P., Petro, L.D., Saha, A., and Shara, M.M., 1996, *Astrophys. J.*, **468**, 241.

Efstathiou, G., and Jones, B.J.T., 1979, *Mon. Not. R. astr. Soc.*, **186**, 133.

Eggen, O.J., Lynden-Bell, D., and Sandage, A.R., 1962, *Astrophys. J.*, **136**, 748.

Elbaz, D., Arnaud, M., and Vangioni-Flam, E., 1995, *Astron. Astrophys.* **303**, 345.

Elson, R.A.W. and Fall, S.M., 1985a, *Publ. Astron. Soc. Pac.*, **97**, 692.

Elson, R.A.W., Fall, S.M., and Freeman, K.C., 1987, *Astrophys. J.*, **323**, 54.

Elson, R.A.W., Gilmore, G.F., and Santiago, B.X., 1995, *Astron. J.*, **110**, 682.

Elson, R., Hut, P., and Inagaki, S., 1987, *Ann. Rev. Astron. Astrophys.*, **25**, 565.

Elson, R.A.W. and Santiago, B.X., 1996a, *Mon. Not. R. astr. Soc.*, **278**, 617.

Elson, R.A.W. and Santiago, B.X., 1996b, *Mon. Not. R. astr. Soc.*, **280**, 971.

Everitt, B.S. and Hand, D.J., 1981, in *Finite Mixture Distributions* (Chapman and Hall, London), p. 30.

Faber, S.M., 1973, *Astrophys. J.*, **179**, 731

Faber, S.M., Dressler, A., Davies, R.L., Burstein, D., Lynden-Bell, D., Terlevich, R., and Wegner, G., 1987, in *Nearly Normal Galaxies*, ed. S.M. Faber (Springer, New York), p. 175.

Faber, S.M. and Gallagher, J.S., 1979, *Ann. Rev. Astron. Astrophys.*, **17**, 135.

Fabian, A.C., 1994, *Ann. Rev. Astron. Astrophys.*, **32**, 277.

Fabian, A.C., Nulsen, P.E.J., and Canizares, C.R., 1984, *Nature*, **310**, 733

Fahlman, G.G., 1993, in *The Globular Cluster–Galaxy Connection*, eds. G.H. Smith and J.P. Brodie (ASP, San Francisco), p. 117.

Fall, S.M., 1979, *Nature*, **281**, 200.

Fall, S.M. and Efstathiou, G., 1980, *Mon. Not. R. astr. Soc.*, **193**, 189.

Fall, S.M. and Rees, M.J., 1977, *Mon. Not. R. astr. Soc.*, **181**, 37p.

Fall, S.M. and Rees, M.J., 1985, *Astrophys. J.*, **298**, 18.

Fall, S.M. and Rees, M.J., 1988, in *Globular Cluster Systems in Galaxies*, eds. J.E. Grindlay and A.G.D. Philip (Reidel, Dordrecht). p. 323.

Federici, L., Bònoli, F., Ciotti, L., Fusi Pecci, F., Marano, B., Lipovetsky, V.A., Neizvestny, S.I., and Spassova, N., 1993, *Astron. Astrophys.*, **274**, 87.

Federici, L., Fusi Pecci, F., and Marano, B., 1990, *Astron. Astrophys.*, **236**, 99.

Ferraro, F.R., Fusi Pecci, F., Cacciari, C., Corsi, C., Buonnano, R., Fahlman, G.G., and Richer, H.B., 1993, *Astron. J.*, **106**, 2324.

Field, G.B. and Saslaw, W.C., 1965, *Astrophys. J.*, **142**, 568.

Fleming, D.E.B., Harris, W.E., Pritchet, C.J., and Hanes, D.A. 1995, *Astron. J.*, **109**, 1044.

Forbes, D.A., Franx, M., Carollo, C.M., and Illingworth, G.D., 1996, *Astrophys. J.*, **467**, 126.

Forte, J.C., Strom, S.E., and Strom, K.M., 1981, *Astrophys. J. Lett.*, **245**, L9.

Freedman, W.L., Madore, B.F., Mould, J.R., Ferrarase, L., Hill, R., Kennicutt, R.C., Jr, *et al.* 1994, *Nature*, **371**, 757.

Freeman, K.C., 1987, *Ann. Rev. Astron. Astrophys.*, **25**, 603.

Freeman, K.C., 1990, in *Dynamics and Interactions of Galaxies*, ed. R. Wielen (Springer, Berlin), p. 36.

Freeman, K.C., Illingworth, G., and Oemler, A., 1983, *Astrophys. J.*, **272**, 488.

Frenk, C.S., and White, S.D.M., 1980, *Mon. Not. R. astr. Soc.*, **193**, 295.

Fritze-v. Alvensleben, U. and Burkert, A., 1995, *Astron. Astrophys.*, **300**, 58.

Fukushige, T. and Heggie, D.C., 1995, *Mon. Not. R. astr. Soc.*, **276**, 206.

Fusi Pecci, F., Battistini, P., Bendinelli, O., Bònoli, F., Cacciari, S., Djorgovski, S.G., Federici, L., Ferraro, F.R., Parmeggiani, G., Weir, N., and Zavatti, F., 1994, *Astron. Astrophys.*, **284**, 349.

Fusi Pecci, F., Bellazzini, M., Cacciari, C., and Ferraro, F., 1995, AJ, **110**, 1664.

Fusi Pecci, F., Buonanno, R., Cacciari, C., Corsi, C.E., Djorgovski, S.G., Federici, L., Ferraro, F.R., Parmeggiani, G., and Rich, R.M., 1996, *Astron. J.*, **112**, 1461.

Fusi Pecci, F., Cacciari, C., Federici, L., and Pasquali, A., 1993a, in *The Globular Cluster–Galaxy Connection*, eds. G.H. Smith and J.P. Brodie (ASP, San Francisco), p. 410.

Fusi Pecci, F., Ferraro, F.R., Bellazzini, M., Djorgovski, S., Piotto, G., and Buonnano, R., 1993b, *Astron. J.*, **105**, 1145.

Fusi Pecci, F., Ferraro, F.R., Corsi, C.E., Cacciari, C., and Buonnano, R., 1992, *Astron. J.*, **104**, 1831.

Gebhardt, K. and Fischer, P., 1995, *Astron. J.*, **109**, 209.

Geisler, D. and Forte, J.C., 1990, *Astrophys. J. Lett.*, **350**, L5.

Geisler, D., Lee, M.G., and Kim, E., 1996, *Astron. J.*, **111**, 1529.

Gilliland, R.L., Edmonds, P.D., Petro, L., Saha, A., and Shara, M.M., 1995, *Astrophys. J.*, **447**, 191.

Gilmore, G., Wyse, R.F.G., and Kuijken, K., 1989, *Ann. Rev. Astron. Astrophys.*, **27**, 555.

Gnedin, O.Y. and Ostriker, J.P., 1997, *Astrophys. J.*, **474**, 223.

Gratton, R. and Ortolani, S., 1988, *Astron. Astrophys. Suppl.*, **73**, 137.

Gott, J.R. and Thuan, T.X., 1976, *Astrophys. J.*, **204**, 649.

Green, E.M. and Norris, J.E., 1990, *Astrophys. J. Lett.*, **353**, L17.

Grillmair, C.J., Faber, S.M., Lauer, T.R., Baum, W.A., Lynds, C.R., and O'Neil, Jr, E.J., 1994a, *Astron. J.*, **108**, 102.

Grillmair, C.J., Freeman, K.C., Bicknell, G.V., Carter, D., Couch, W.J., Sommer-Larsen, J., and Taylor, K. 1994b, *Astrophys. J. Lett.*, **422**, L9.

Grillmair, C.J., Freeman, K.C., Irwin, M., and Quinn, P.J., 1995, *Astron. J.*, **109**, 2553.

Grillmair, C.J., Ajhar, E.A., Faber, S.M., Baum, W.A., Holtzman, J.A., Lauer, T.R., Lynds, C.R., and O'Neil, Jr, E.J., 1996, *Astron. J.*, **111**, 2293.

Grindlay, J.E, 1993, in *The Globular Cluster–Galaxy Connection*, eds. G.H. Smith and J.P. Brodie (ASP, San Francisco), p. 156.

Gunn, J.E., 1980, in *Globular Clusters*, eds. D. Hanes and B. Madore (Cambridge University Press), p. 301.

Gunn, J.E. and Griffin, R.F., 1979, *Astron. J.*, **84**, 752.

Haiman, Z., Rees, M.J., and Loeb, A., 1996, *Astrophys. J.*, **467**, 522.

Hanes, D.A., 1977, *Mem. R. astr. Soc.*, **84**, 45.

Hanes, D.A. and Harris, W.E., 1986, *Astrophys. J.*, **304**, 599.

Hanes, D.A. and Whittaker, D.G., 1987, *Astron. J.*, **94**, 906.

Harris, G.L.H., Geisler, D., Harris, H.C., and Hesser. J.E., 1992, *Astron. J.*, **104**, 613.

Harris, H.C., Harris, G.L.H., and Hesser, J.E., 1988, in *Globular Cluster Systems in Galaxies*, eds. J.E. Grindlay and A.G.D. Phillip (Reidel, Dordrecht), p. 205.

Harris, H.C. and Harris, W.E., 1997, in *Astrophysical Quantities*, 4th edition, ed. A.N. Cox, in press.

Harris, W.E., 1976, *Astron. J.*, **81**, 1095.

Harris, W.E., 1986, *Astron. J.*, **91**, 822.

Harris, W.E., 1991, *Ann. Rev. Astron. Astrophys.*, **29**, 543.

Harris, W.E., 1996, *Astron. J.*, **112**, 487.

Harris, W.E., Allwright, J.W.B., Pritchet, C.J., and van den Bergh, S., 1991, *Astrophys. J. Suppl.*, **76**, 115.

Harris, W.E., Harris, H.C., and Harris, G.L.H., 1984, *Astron. J.*, **89**, 216.

Harris, W.E., Pritchet, C.J., and McClure, R.D., 1995, *Astrophys. J.*, **441**, 120.

Harris, W.E. and Pudritz, R.E., 1994, *Astrophys. J.*, **429**, 177.

Harris, W.E. and Racine, R., 1979, *Ann. Rev. Astron. Astrophys.*, **17**, 241.

Harris, W.E. and van den Bergh, S., 1981, *Astron. J.*, **86**, 1627.

Hartwick, F.D.A., 1987, in *The Galaxy*, eds. G. Gilmore and B. Carswell (Reidel, Dordrecht), p. 281.

Haynes, R.F. and Milne, D.K. (eds.), 1991, *The Magellanic Clouds* (Dordrecht, Kluwer).

Heavens, A., 1988, *Mon. Not. R. astr. Soc.*, **232**, 339.

Heggie, D.C., and Hut, P., 1996, in *IAU Symposium 174, Dynamical Evolution of Star Clusters –
   Confrontation of Theory and Observation*, eds. P.Hut and J. Makino (Kluwer, Dordrecht), p. 303.

Hertz, P. and Grindlay, J.E., 1983a, *Astrophys. J. Lett.*, **267**, L83.

Hertz, P. and Grindlay, J.E., 1983b, *Astrophys. J.*, **275**, 105.

Hibbard, J.E. and van Gorkom, J.H., 1996, *Astron. J.*, **111**, 655

Hilker, M. and Kissler-Patig, M. 1996, *Astron. Astrophys.*, **314**, 357.

Hills, J.G. and Day, C.A., 1976, *Astrophys. Lett.*, **17**, 87.

Ho, L.C. and Filippenko, A.V., 1996a, *Astrophys. J. Lett.*, **466**, L83.

Ho, L.C. and Filippenko, A.V., 1996b, *Astrophys. J.*, **427**, 600.

Hodge, P.W., 1961, *Astron. J.*, **66**, 83.

Hodge, P.W., 1965, *Astrophys. J.*, **141**, 308.

Hodge, P.W., 1987, *Publ. Astron. Soc. Pac.*, **99**, 724.

Hoyle, F. and Schwarzschild, M., 1955, *Astrophys. J. Suppl.*, **2**, 1.

Hopp, U., Wagner, S.J., and Richtler, T., 1995, *Astron. Astrophys.*, **296**, 633.

Holtzman, J.A., Faber, S.M., Shaya, E.J., Lauer, T.R., Groth, E.J., Hunter, D.A., *et al.*, 1992, *Astron.
   J.*, **103**, 691.

Holtzman, J.A., Watson, A.M., Mould, J.R., Gallagher, J.S., Ballester, G.E., Burrows, C.J., *et al.*,
   1996, *Astron. J.*, **112**, 416.

Hubble, E., 1932, *Astrophys. J.*, **76**, 44.

Huchra, J.P., 1993, in *The Globular Cluster–Galaxy Connection*, eds. G.H. Smith and J.P. Brodie
   (ASP, San Francisco), p. 420.

Huchra, J.P., Brodie, J.P., and Kent, S.M., 1991, *Astrophys. J.*, **370**, 495.

Huchra, J.P., Stauffer, J., and van Speybroeck, L., 1982, *Astrophys. J. Lett.*, **259**, L57.

Hui, X., Ford, H.C., Freeman, K.C., and Dopita, M.A., 1995, *Astrophys. J.*, **449**, 592.

Hunter, C. and Tremaine, S., 1977, *Astron. J.*, **82**, 262.

Hut, P. and Djorgovski, S., 1992, *Nature*, **359**, 806.

Ibata, R., Gilmore, G.F., and Irwin, M.J., 1994, *Nature*, **370**, 194.

Ibata, R., Gilmore, G.F., and Irwin, M.J., 1995, *Mon. Not. R. astr. Soc.*, **277**, 781.

Iben, I., Jr, 1974, *Ann. Rev. Astron. Astrophys.*, **12**, 215.

Iben, I., Jr, 1986, *Mem. Soc. Astron. Ital.*, **57**, 453.

Iben, I., Jr, and Faulkner, J., 1968, *Astrophys. J.*, **153**, 101.

Iben, I., Jr, and Renzini, A., 1983, *Ann. Rev. Astron. Astrophys.*, **21**, 271.

Illingworth, G. 1981, in *The Structure and Evolution of Normal Galaxies*, eds. S.M. Fall and D.
   Lynden-Bell (Cambridge University Press), p. 27.

Irwin, M.J. and Hatzidimitriou, D., 1993 in *The Globular Cluster–Galaxy Connection*, eds. G.H.
   Smith and J.P. Brodie (ASP, San Francisco), p. 322.

Jacoby, G.H., *et al.*, 1992, *Publ. Astron. Soc. Pac.*, **104**, 599.

Jimenez., R. and Padoan, P., 1996, *Astrophys. J. Lett.*, **463**, L17.

Jimenez, R., Thejll, P., Jorgensen, U.G., MacDonald, J., and Pagel, B., 1996, *Mon. Not. R. astr. Soc.*,
   282, 926

Jog, C.J. and Solomon, P.M., 1992, *Astrophys. J.*, **387**, 152.

Johnson, H.R. and Zuckerman, B. (eds.), 1989, IAU Colloquium No. 106, *The Evolution of Peculiar
   Red Giant Stars* (Cambridge University Press).

Jones, R.V., Carney, B.W., Storm, J., and Latham, D.W., 1992, *Astrophys. J.*, **386**, 646.

Judge, P.G. and Stencel, R.E., 1991, *Astrophys. J.*, **371**, 357.

Kaisler, D., Harris, W.E., Crabtree, D.R., and Richer, H.B. 1996, *Astron. J.*, **111**, 2224.

Kang, H., Shapiro, P.R., Fall, S.M., and Rees, M.J., 1990, *Astrophys. J.*, **363**, 488.

Kauffmann, G., 1996, *Mon. Not. R. astr. Soc.*, **281**, 487.

Kauffmann, G., Guiderdoni, B., and White, S.D.M., 1994, *Mon. Not. R. astr. Soc.*, **267**, 981.

Kennicutt, R.C. and Chu, Y.-H., 1988, *Astron. J.*, **95**, 720.

King, I.R., 1962, *Astron. J.*, **67**, 471.

King, I.R., 1966, *Astron. J.*, **71**, 64.

Kinman, T.D., 1959, *Mon. Not. R. astr. Soc.*, **119**, 559.

Kissler, M., Richtler, T., Held, E.V., Grebel, E.K, Wagner, S., and Capaccioli, M. 1994, *Astron. Astrophys.*, 287, 463.

Kissler-Patig, M., Richtler, T., and Hilker, M., 1996, *Astron. Astrophys.*, **308**, 704.

Kissler-Patig, M., Kohle, S., Hilker, M., Richtler, T., Infante, L., and Quintana, H., 1997, *Astron. Astrophys.*, **319**, 470.

Kohle, S., Kissler-Patig, M., Richtler, T., Infante, L., and Quintana, H., 1996, *Astron. Astrophys.*, **309**, L39.

Kormendy, J. and Djorgovski, S.G., 1989, *Ann. Rev. Astron. Astrophys.*, **27**, 235

Kormendy, J. and Westphal, D.J., 1989, *Astrophys. J.*, **338**, 752.

Kraft, R.P., Sneden, C., Langer, G.E., and Shetrone, M.D., 1993, *Astron. J.*, **106**, 1490.

Kumai, Y., Basu, B., and Fujimoto, M., 1993a, *Astrophys. J.*, **404**, 144.

Kumai, Y., Hashi, Y., and Fujimoto, M., 1993b, *Astrophys. J.*, **416**, 576.

Kundić, T. and Ostriker, J.P., 1995, *Astrophys. J.*, **438**, 702.

Kwan, J., 1979, *Astrophys. J.*, **229**, 567.

Kwan, J. and Valdes, F., 1983, *Astrophys. J.*, **271**, 604.

Laird, J.B., Carney, B.W., and Latham, D.E., 1993, in *The Globular Cluster–Galaxy Connection*, eds. G.H. Smith and J.P. Brodie (ASP, San Francisco), p. 95.

Langer, G.E., Bolte, M., Prosser, C.F., and Sneden, C., 1993 in *The Globular Cluster–Galaxy Connection*, eds. G.H. Smith and J.P. Brodie (ASP, San Francisco), p. 206.

Langer, G.E. and Hoffman, R.D., 1995, *Publ. Astron. Soc. Pac.*, **107**, 1177.

Larson, R.B., 1969a, *Mon. Not. R. astr. Soc.*, **145**, 271.

Larson, R.B., 1969b, *Mon. Not. R. astr. Soc.*, **145**, 405.

Larson, R.B., 1974a, *Mon. Not. R. astr. Soc.*, **166**, 585.

Larson, R.B., 1974b, *Mon. Not. R. astr. Soc.*, **169**, 229.

Larson, R.B., 1975, *Mon. Not. R. astr. Soc.*, **173**, 671.

Larson, R.B., 1976, *Mon. Not. R. astr. Soc.*, **176**, 31.

Larson, R.B., 1978, in *Infrared Astronomy*, eds. G. Setti and G.G. Fazio (Reidel, Dordrecht), p. 137.

Larson, R.B., 1988, in *Globular Cluster Systems in Galaxies*, eds. J.E. Grindlay and A.G.D. Phillip (Reidel, Dordrecht), p. 311.

Larson, R.B., 1990, in *Physical Processes in Fragmentation and Star Formation*, eds. R. Capuzzo-Dolcetta, C. Chiosi, and A. Di Fazio (Kluwer, Dordrecht), p. 389.

Larson, R.B., 1993, in *The Globular Cluster-Galaxy Connection*, eds. G.H. Smith and J.P. Brodie (ASP, San Francisco), p. 675.

Larson, R.B., 1996, in *Formation of the Galactic Halo...Inside and Out*, eds. H. Morrison and A. Sarajedini (ASP, San Francisco), p.241.

Lauer, T.R., Ajhar, E.A., Byun, Y.I., Dressler, A., Faber, S.M., Grillmair, C., Kormendy, J., Richstone, D., and Tremaine, S., 1995, *Astron. J.*, **110**, 2622.

Lee, H.M. and Goodman, J., 1995, *Astrophys. J.*, **443**, 109.

Lee, M.G. and Geisler, D., 1993, *Astron. J.*, **106**, 493.

Lee, S., Schramm, D.N., and Mathews, G.J., 1995, *Astrophys. J.*, **449**, 616.

Lee, Y.-W., 1990, *Astrophys. J.*, **363**, 159.

Lee, Y.-W., 1993, in *The Globular Cluster-Galaxy Connection*, eds. G.H. Smith and J.P. Brodie (ASP, San Francisco), p. 142.

Lee, Y.-W., Demarque, P., and Zinn, R., 1994, *Astrophys. J.*, **423**, 248.

Lehnert, M.D., Bell, R.A., and Cohen, J.G., 1991, *Astrophys. J.*, **367**, 514.

Leonard, P.J.T., 1989, *Astron. J.*, **98**, 217.

Leonard, P.J.T., 1996a, in *The Origins, Evolution and Destinies of Binary Stars in Clusters*, eds. E.F. Milone and J.-C. Mermilliod (ASP, San Francisco), p. 337.

Leonard, P.J.T., 1996b, *Astrophys. J.*, **470**, 521.

Leonard, P.J.T. and Fahlman, G.G., 1991, *Astron. J.*, **102**, 994.

Leonard, P.J.T., Richer, H.B., and Fahlman, G.G., 1992, *Astron. J.*, **104**, 2104.

Lightman, A.L. and Shapiro, S.L., 1978, *Rev. Mod. Phys.*, **50**, 437.

Lin, D.N.C. and Murray, S.D., 1992, *Astrophys. J.*, **394**, 523.

Lloyd Evans, T., 1975, *Mon. Not. R. astr. Soc.*, **171**, 647.

Lombardi, J.C., Jr, Rasio, F.A., and Shapiro, S.L., 1996, *Astrophys. J.*, **468**, 797.

Long, K., Ostriker, J.P., and Aguilar, L., 1992, *Astrophys. J.*, **103**, 703.

Lugger, P.M., Cohn, H., Grindlay, J.E., Bailyn, C.D., and Hertz, P., 1987, *Astrophys. J.*, **320**, 482.

Lupton, R.H., 1989, *Astron. J.*, **97**, 1350.

Lutz, D., 1991, *Astron. Astrophys.*, **245**, 31.

Lynden-Bell, D., 1967, *Mon. Not. R. astr. Soc.*, **136**, 101.

Lynden-Bell, D., 1982, *The Observatory*, **102**, 202.

Lynden-Bell, D., 1994, in *Dwarf Galaxies*, eds. G. Meylan and P. Prugniel (ESO, Munich), p. 589.

Lynden-Bell, D., and Lynden-Bell, R.M., 1995, *Mon. Not. R. astr. Soc.*, **275**, 429.

MacLow, M.-M. and Shull, J.M., 1986, *Astrophys. J.*, **302**, 585.

Majewski, S.R., 1992, *Astrophys. J. Suppl.*, **78**, 87.

Majewski, S.R., 1993, *Ann. Rev. Astron. Astrophys.*, **31**, 575.

Majewski, S.R., 1994a, *Astrophys. J. Lett.*, **431**, L17.

Majewski, S.R., 1994b, in *Astronomy from Wide-Field Imaging, IAU Symposium 161*, eds. H.T. MacGillivray *et al.* (Kluwer, Dordrecht), p. 425.

Mandushev, G., Spassova, N., and Staneva, A., 1991, *Astron. Astrophys.*, **252**, 94.

Mateo, M., 1993, in *The Globular Cluster–Galaxy Connection*, eds. G.H. Smith and J.P. Brodie (ASP, San Francisco), p. 387.

Mateo, M., Harris, H.C., Nemec, J., and Olszewski, E.W., 1990, *Astron. J.*, **100**, 469.

Mateo, M., Hodge, P.W., and Schommer, R.A., 1986, *Astrophys. J.*, **311**, 113.

Mathews, G.J. and Schramm, D.N., 1993, *Astrophys. J.*, **404**, 468.

Mathewson, D.H., Cleary, M.N., and Murray, J.D., 1974, *Astrophys. J.*, **190**, 291.

Matteucci, F., 1994, *Astron. Astrophys.*, **288**, 57.

Matteucci, F. and Tornambè, A., 1987, *Astron. Astrophys.*, **185**, 51.

Matteucci, F. and Tornambè, A., 1994, *Astron. Astrophys.*, **288**, 57.

Mayor, M., Duquennoy, A., Alimenti, A., Andersen, J., and Nördstrom, N., 1996, in *The Origins, Evolution, and Destinies of Binary Stars in Clusters*, eds. G. Milone and J.-C. Mermilliod (ASP, San Francisco), p. 190.

McClure, R.D., VandenBerg, D.A., Smith, G.H., Fahlman, G.G., Richer, H.B., Hesser, J.E., Harris, W.E., Stetson, P.B., and Bell, R.A., 1986, *Astrophys. J. Lett.*, **307**, L49.

McCrea, W.H., 1964, *Mon. Not. R. astr. Soc.*, **128**, 147.

McCrea, W.H., 1982, in *Progress in Cosmology*, ed. A.W. Wolfendale (Reidel, Dordrecht), p. 239.

McKee, C.F., Zweibel, E.G., Goodman, A.A., and Heiles, C., 1993, in *Protostars and Planets III*, eds. E.H. Levy and J.I. Lunine (University of Arizona Press, Tucson), p. 327.

McLachlan, G.J., and Basford, K.E., 1988, in *Mixture Models: Inference and Applications to Clustering* (Marcel Dekker, New York).

McLaughlin, D.E., 1994, *Publ. Astron. Soc. Pac.*, **106**, 47.

McLaughlin, D.E., 1995, *Astron. J.*, **109**, 2034.

McLaughlin, D.E., Harris, W.E., and Hanes, D.A., 1994, *Astrophys. J.*, **422**, 486.

McLaughlin, D.E. and Pudritz, R., 1996, *Astrophys. J.*, **457**, 578.

Melnick, J., Moles, M., and Terlevich, R., 1985, *Astron. Astrophys.*, **149**, L24.

Merritt, D., 1993a, in *Structure and Dynamics of Star Clusters*, eds. D. Djorgovski and G. Meylan (ASP, San Francisco), p. 117.

Merritt, D., 1993b, in *Structure, Dynamics and Chemical Evolution of Elliptical Galaxies*, eds. I.J. Danziger, W.W. Zeilinger and K. Kjär (ESO, Munich), p. 275.

Merritt, D. and Saha, P., 1993, *Astrophys. J.*, **409**, 75.

Merritt, D. and Tremblay, B., 1994, *Astron. J.*, **108**, 514.

Meurer, G.R., 1995, *Nature*, **375**, 472.

Meurer, G.R., Freeman, K.C., Dopita, M.A., and Cacciari, C., 1992, *Astron. J.*, **103**, 60.

Meurer, G.R., Heckman, T.M., Leitherer, C., Kinney, A., and Robert, C., 1995, *Astron. J.*, **110**, 2665.

Meylan, G., 1987, *Astron. Astrophys.*, **184**, 144.

Meylan, G., 1988, *Astron. Astrophys.*, **191**, 215.

Meylan, G., 1989, *Astron. Astrophys.*, **214**, 106.

Meylan, G. and Djorgovski, S.G., 1987, *Astrophys. J. Lett.*, **322**, L91.

Meylan, G., Dubath, P., and Mayor, M., 1991, in *The Magellanic Clouds, IAU Symposium No. 148*, eds. R.F. Haynes and D.K. Milne (Kluwer, Dordrecht), p. 211.

Meylan, G., Dubath, P., Mayor, M., and Magain, P., 1989, *ESO Mess.*, **55**, 55.

Meylan, G., and Heggie, D.C., 1996, *Astron. Astrophys. Rev.*, in press.

Meylan, G., and Pryor, C., 1993, in *Structure, Dynamics and Chemical Evolution of Elliptical Galaxies*, eds. I.J. Danziger, W.W. Zeilinger and K. Kjär (ESO, Munich), p. 31.

Michie, R.W., 1963, *Mon. Not. R. astr. Soc.*, **125**, 127.

Mighell, K.J., Rich, R.M., Shara, M., and Fall, S.M., 1996, *Astron. J.*, **111**, 2314.

Mihalas, D. and Binney, J., 1981, *Galactic Astronomy* (Freeman, San Francisco).

Miller, G.E. and Scalo, J.M., 1979, *Astrophys. J. Suppl.*, **41**, 513.

Minniti, D., 1995, *Astron. J.*, **109**, 1663.

Minniti, D., Meylan, G., and Kissler-Patig, M., 1996, *Astron. Astrophys.*, **312**, 49.

Moore, B., 1996, *Astrophys. J. Lett.*, **416**, L13.

Morgan, S. and Lake, G., 1989, *Astrophys. J.*, **339**, 171.

Morgan, W.W., 1959, *Astron. J.*, **64**, 432.

Morrison, H. and Sarajedini, A. (eds.), 1996, *Formation of the Galactic Halo...Inside and Out* (ASP, San Francisco).

Mould, J.R., Oke, J.B., de Zeeuw, P.T., and Nemec, J.M., 1990, *Astron. J.*, **99**, 1823.

Mukherjee, K., Anthony-Twarog, B.J., and Twarog, B.A., 1992, *Publ. Astron. Soc. Pac.*, **104**, 561.

Murali, C. and Weinberg, M.D., 1996, *Mon. Not. R. astr. Soc.*, **288**, 767.

Murphy, B.W., Cohn, H.N., and Hut, P., 1990, *Mon. Not. R. astr. Soc.*, **245**, 355.

Murray, S.D. and Lin, D.N.C., 1990, *Astrophys. J.*, **357**, 105.

Murray, S.D. and Lin, D.N.C., 1992, *Astrophys. J.*, **400**, 265.

Murray, S.D. and Lin, D.N.C., 1993, in *The Globular Cluster–Galaxy Connection*, eds. G.H. Smith and J.P. Brodie (ASP, San Francisco), p. 738.

Murtagh, F. and Heck, A., 1987, *Multivariate Data Analysis* (Reidel, Dordrecht).

Niss, B., Jorgensen, H.E., and Lautsen, S., 1978, *Astron. Astrophys. Suppl.*, **32**, 387.

Norris, J.E., 1980, in *Globular Clusters*, eds. D. Hanes and B. Madore (Cambridge University Press), p. 113.

Norris, J.E., 1986, *Astrophys. J. Suppl.*, **61**, 667.

Norris, J.E., Da Costa, G.S., amd Mighell, K.J., 1996, *Astrophys. J.*, **462**, 241.

Norris, J.E. and Freeman, K.C., 1983, *Astrophys. J.*, **266**, 130.

Norris, J.E. and Ryan, S.G., 1989, *Astrophys. J.*, **340**, 739.

Norris, J.E. and Smith, G.H., 1981, in *Astrophysical Parameters for Globular Clusters*, eds. A.G.D. Philip and D.S. Hayes (L. Davis Press, Schenectady), p. 109.

O'Connell, R.W., Gallagher, J.S., and Hunter, D.A., 1994, *Astrophys. J.*, **433**, 65.

O'Connell, R.W., Gallagher, J.S., Hunter, D.A., and Colley, W.N., 1995, *Astrophys. J. Lett.*, **446**, L1.

Oh, K.S. and Lin, D.N.C., 1992, *Astrophys. J.*, **386**, 519.

Oh, K.S., Lin, D.N.C., and Aarseth, S.J., 1992, *Astrophys. J.*, **386**, 506.

Okazaki, T. and Tosa, M., 1995, *Mon. Not. R. astr. Soc.*, **274**, 48.

Olszewski, E.W., 1993, in *The Globular Cluster–Galaxy Connection*, eds. G.H. Smith and J.P. Brodie (ASP, San Francisco), p. 351.

Olszewski, E.W., Schommer, R.A., Suntzeff, N.B., and Harris, H.C., 1991, *Astron. J.*, **101**, 515.

Oosterhoff, P.T., 1939, *The Observatory*, **62**, 104.

Oosterhoff, P.T., 1944, *Bull. Astron. Inst. Neth.*, **10**, 55.

Ostriker, J.P., Binney, J., and Saha, P., 1989, *Mon. Not. R. astr. Soc.*, **241**, 849.

Ostriker, J.P., Spitzer, L., and Chevalier, R.A., 1972, *Astrophys. J. Lett.*, **176**, L51.

Ostrov, P., Geisler, D., and Forte, J.C., 1993, *Astron. J.*, **105**, 1762.

Paresce, F. and De Marchi, G., 1994, *Astrophys. J.*, **427**, L33.

Paresce, F., De Marchi, G., and Romaniello, M., 1995, *Astrophys. J.*, **440**, 216.

Peebles, P.J.E., 1969, *Astrophys. J.*, **155**, 393.

Peebles, P.J.E., 1984, ApJ, **277**, 470.

Peebles, P.J.E., 1989, in *The Epoch of Galaxy Formation*, eds. C.S. Frenk, R.S. Ellis, T. Shanks, A. Heavens and J.A. Peacock (Kluwer, Dordrecht), p. 1.

Peebles, P.J.E. and Dicke, R.H., 1968, *Astrophys. J.*, **154**, 891.

Perelmuter, J.-M. 1995, *Astrophys. J.*, **454**, 762.

Perelmuter, J.-M., Brodie, J.P., and Huchra, J.P., 1995, *Astron. J.*, **110**, 620.

Perelmuter, J.-M. and Racine, R., 1995, *Astron. J.*, **109**, 1055.

Peterson, C.J., 1974, *Astrophys. J. Lett.*, **190**, L17.

Phinney, E.S., 1993, in *Structure and Dynamics of Star Clusters*, eds. S.G. Djorgovski and G. Meylan (ASP, San Francisco), p. 141.

Phinney, E.S. and Kulkarni, S.R., 1994, *Ann. Rev. Astron. Astrophys.*, **32**, 591.

Piotto, G., 1991, in *The Formation and Evolution of Star Clusters*, ed. K. Janes (ASP, San Francisco), p. 200.

Piotto, G., Cool, A.M., and King, I.R., 1997, *Astron. J.*, **113**, 1345.

Preston, G.W., Beers, T.C., and Shectman, S.A., 1994, *Astron. J.*, **108**, 538.

Preston, G.W., Shectman, S.A., and Beers, T.C., 1991, *Astrophys. J.*, **375**, 121.

Pritchet, C.J. and van den Bergh, S., 1994, *Astron. J.*, **107**, 1730.

Proffitt, C.R. and VandenBerg, D.A., 1991, *Astrophys. J. Suppl.*, **77**, 473.

Pryor, C., McClure, R.D., Fletcher, J.M., Hartwick, F.D.A., and Kormendy J., 1986, *Astron. J.*, **91**, 546.

Pryor, C., McClure, R.D., Fletcher, J.M., and Hesser, J.E., 1989, *Astron. J.*, **98**, 596.

Pryor, C. and Meylan, G., 1993, in *Structure, Dynamics and Chemical Evolution of Elliptical Galaxies*, eds. I.J. Danziger, W.W. Zeilinger and K. Kjär (ESO, Munich), p. 357.

Quinn, P.J., Hernquist, L., and Fullagar, D.P., 1993, *Astrophys. J.*, **403**, 74.

Racine, R. 1968, *Publ. Astron. Soc. Pac.*, **80**, 326.

Racine, R., 1980, in *Star Clusters*, ed. J.E. Hesser (Reidel, Dordrecht), p. 369.

Racine, R., 1991, *Astron. J.*, **101**, 865.

Ratnatunga, K.U. and Freeman, K.C., 1989, *Astrophys. J.*, **339**, 126.

Reed, L.G., Harris, G.L.H., and Harris, W.E., 1994, *Astron. J.*, **107**, 555.

Rees, R.F. and Cudworth, K.M. 1991, *Astron. J.*, **102**, 152.

Reid, N., 1990, *Mon. Not. R. astr. Soc.*, **247**, 70.

Reiz, A., 1954, *Astrophys. J.*, **120**, 351.

Renzini, A., 1977, in *Advanced Stages of Stellar Evolution*, eds. P. Bouvier and A. Maeder (Geneva Observatory, Sauverny), p. 151.

Renzini, A., and Fusi Pecci, F., 1988, *Ann. Rev. Astron. Astrophys.*, **26**, 199.

Renzini, A., Bragaglia, A., Ferraro, F.R., Gilmozzi, R., Ortolani, S., Holberg, J.B., Liebert, J., Wesemael, F., and Bohlin, R.C., 1996, *Astrophys. J. Lett.*, **465**, L23.

Rich, R.M., 1996, in *Formation of the Galactic Halo...Inside and Out*, eds. H. Morrison and A. Sarajedini (ASP, San Francisco), p. 24.

Rich, R.M., Mighell, K.J., Freedman, W.L., and Neill, J.D., 1996, *Astron. J.*, **111**, 768.

Richer, H.B., Crabtree, D.R., Fabian, A.C., and Lin, D.N.C. 1993, *Astron. J.*, **105**, 877.

Richer, H.B. and Fahlman, G.G., 1992, *Nature*, **358**, 383.

Richer, H.B., Fahlman, G.G., Ibata, R.A., Stetson, P.B., Bell, R.A., Bolte, M., *et al.*, 1995, *Astrophys. J. Lett.*, **451**, L17.

Richer, H.B., Harris, W.E., Fahlman, G.G., Bell, R.A., Bond. H.E., Hesser, J.E., *et al.*, 1996, *Astrophys. J.*, **463**, 602.

Richstone, D.O. and Tremaine, S., 1986, *Astron. J.*, **92**, 72.

Richtler, T., 1993, in *The Globular Cluster–Galaxy Connection*, eds. G.H. Smith and J.P. Brodie (ASP, San Francisco), p. 375.

Richtler, T. and Fichtner, H., 1993, in *The Globular Cluster–Galaxy Connection*, eds. G.H. Smith and J.P. Brodie (ASP, San Francisco), p. 725.

Richtler, T., 1995, in *Reviews in Modern Astronomy*, vol. 8, ed. G. Klare (Springer, Berlin), p. 163.

Roberts, M.S., 1960, *Astron. J.*, **65**, 457.

Rodgers, A.W. and Paltoglou, G., 1984, *Astrophys. J. Lett.*, **283**, L5.

Rodgers, A.W. and Roberts, W.H., 1994, *Astron. J.*, **107**, 1737.

Rood, R.T., 1973, *Astrophys. J.*, **184**, 815.

Rood, H.J., Page, T.L., Kintner, E.C., and King, I.R., 1972, *Astrophys. J.*, **179**, 627.

Rosenblatt, E.I., Faber, S.M., and Blumenthal, G.R., 1988, *Astrophys. J.*, **330**, 191.

Rubenstein, E.P. and Bailyn, C.D., 1996, *Astron. J.*, **111**, 260.

Rubenstein, E.P. and Bailyn, C.D., 1997, *Astrophys. J.*, **474**, 701.

Saffer, R.A. (ed.), 1993, *Blue Stragglers* (ASP, San Francisco).

Saha, A., 1985, *Astrophys. J.*, **289**, 310.

Salpeter, E.E., 1955, *Astrophys. J.*, **121**, 161.

Sandage, A., 1953, *Astron. J.*, **58**, 61.

Sandage, A., 1961, in *The Hubble Atlas of Galaxies* (Carnegie Institute, Washington).

Sandage, A. 1982, *Astrophys. J.*, **252**, 553.

Sandage, A., 1990, *J. R. astr. Soc. Can.*, **84**, 70.

Sandage, A., 1993a, *Astron. J.*, **106**, 687.

Sandage, A., 1993b, *Astron. J.*, **106**, 719.

Sandage, A. and Fouts, G., 1987, *Astron. J.*, **93**, 592.

Sandage, A., Freeman, K.C., and Stokes, N.R., 1970, *Astrophys. J.*, **160**, 831.

Sandage A. and Schwarzschild, M., 1952, *Astrophys. J.*, **116**, 463.

Sandage, A. and Smith, L.L., 1966, *Astrophys. J.*, **144**, 886.

Sandage, A. and Tammann, G.A., 1995, *Astrophys. J.*, **446**, 1.

Sandage, A. and Visvanathan, N., 1978, *Astrophys. J.*, **225**, 742.

Sandage, A. and Wallerstein, G., 1960, *Astrophys. J.*, **131**, 598.

Santiago, B.X. and Djorgovski, S.G., 1993, *Mon. Not. R. astr. Soc.*, **261**, 753.

Sarajedini, A., 1993, in *Blue Stragglers*, ed. R.A. Saffer (ASP, San Francisco), p. 14.

Sarajedini, A. and Demarque, P., 1990, *Astrophys. J.*, **365**, 219.

Sarajedini, A. and Geisler, D., 1996, *Astron. J.*, **112**, 2013.

Scalo, J.M., 1986, *Fundam. Cosmic Phys.*, **11**, 1.

Schommer, R.A., 1993, in *The Globular Cluster–Galaxy Connection*, eds. G.H. Smith and J.P. Brodie (ASP, San Francisco), p. 458.

Schommer, R.A., Christian, C.A., Caldwell, N., Bothun, G.D., and Huchra, J.P., 1991, *Astron. J.*, **101**, 873.

Schommer, R.A., Olszewski, E.W., Suntzeff, N.B., and Harris, H.C., 1992, *Astron. J.*, **103**, 447.

Schramm, D.N., Dearborn, D.S.P., and Truran, J.W., 1995, *Comments on Astrophysics*, **17**, 343.

Schuster, W.J., Parrao, L., and Contreras Martinez, M.E., 1993, *Astron. Astrophys. Suppl.*, **97**, 951.

Schwarzschild, M., Searle, L., and Howard, R., 1955, *Astrophys. J.*, **122**, 353.

Schweizer, F., 1982, *Astrophys. J.*, **252**, 455.

Schweizer, F., 1987, in *Nearly Normal Galaxies*, ed. S.M. Faber (Springer, New York), p. 18.

Schweizer, F., 1990, in *Dynamics and Interactions of Galaxies*, ed. R. Wielen (Springer, Berlin), p. 60.

Schweizer, F., Miller, B.W., Whitmore, B.C., and Fall, S.M., 1996, *Astron. J.*, **112**, 1839.

Schweizer, F. and Seitzer, P., 1993, *Astrophys. J. Lett.*, **417**, L29.

Searle, L. and Zinn, R., 1978, *Astrophys. J.*, **225**, 357.

Secker, J., 1992, *Astron. J.*, **104**, 1472.

Secker, J., Geisler, D., McLaughlin, D., and Harris, W.E., 1995, *Astron. J.*, **109**, 1019.

Secker, J. and Harris, W.E., 1993, *Astron. J.*, **105**, 1358.

Shapiro, P.R. and Kang, H., 1987, *Astrophys. J.*, **318**, 32.

Shetrone, M.D., 1996, *Astron. J.*, **112**, 1517.

Shi, X., 1995, *Astrophys. J.*, **446**, 637.

Silk, J. and Wyse, R.F.G., 1993, *Physics Reports*, **231**, 293.

Smith, E.O., Neill, J.D., Mighell, K.J., and Rich, R.M., 1996, *Astron. J.*, **111**, 1596.

Smith, G.H., 1987, *Publ. Astron. Soc. Pac.*, **99**, 67.

Smith, G.H. and McClure, R.D., 1987, *Astrophys. J.*, **316**, 206.

Sneden, C., Kraft, R.P., Prosser, C.F., and Langer, G.E., 1992, *Astron. J.*, **104**, 2121.

Spassova, N.M., Staneva, A.V., and Golev, V.K., 1988, in *Globular Cluster Systems in Galaxies*, eds. J.E. Grindlay and A.G.D. Phillip (Reidel, Dordrecht), p. 569.

Spinrad, H. and Schweizer, F., 1972, *Astrophys. J.*, **171**, 403.

Spitzer, L., 1958, *Astrophys. J.*, **127**, 17.

Spitzer, L., 1975, in *Dynamics of Stellar Systems*, ed. A. Hayli (Reidel, Dordrecht), p. 3.

Spitzer, L., 1987, *Dynamical Evolution of Globular Clusters* (Princeton University Press, Princeton).

Spitzer, L. and Chevalier, R.A., 1973, *Astrophys. J.*, **183**, 565.

Spitzer, L. and Thuan, T.X., 1972, *Astrophys. J.*, **175**, 31.

Stetson, P.B., 1990, *Publ. Astron. Soc. Pac.*, **102**, 932.

Stetson, P.B., 1993, in *The Globular Cluster–Galaxy Connection*, eds. G.H. Smith and J.P. Brodie (ASP, San Francisco), p. 14.

Stetson, P.B., VandenBerg, D.A., and Bolte, M., 1996, *Publ. Astron. Soc. Pac.*, **108**, 560.

Stetson, P.B., VandenBerg, D.A., Bolte, M., Hesser, J.H., and Smith, G.H., 1989, *Astron. J.*, **97**, 1360.

Stiavelli, M., Piotto, G.P., Capaccioli, M., and Ortolani, S., 1991, in *The Formation and Evolution of Star Clusters*, ed. K. Janes (ASP, San Francisco), p. 449.

Stiavelli, M., Piotto, G.P., and Capaccioli, M., 1992, in *Morphological and Physical Classification of Galaxies*, eds. G. Longo, M. Capaccioli and G. Busarello (Kluwer, Dordrecht), p. 455.

Straniero, O. and Chieffi, A., 1991, *Astrophys. J. Suppl.*, **76**, 525.

Strom, S.E., Forte, J.C., Harris, W.E., Strom, K.M., Wells, D.C., Smith, M.G., 1981, *Astrophys. J.*, **245**, 416.

Stryker, L.L., 1993, *Publ. Astron. Soc. Pac.*, **105**, 1081.

Suntzeff, N., 1989, in *The Abundance Spread Within Globular Clusters: Spectroscopy of Individual Stars*, eds. G. Cayrel de Strobel, M. Spite and T. Lloyd Evans (Observatoire de Paris), p. 71.

Suntzeff, N., 1993, in *The Globular Cluster–Galaxy Connection*, eds. G.H. Smith and J.P. Brodie (ASP, San Francisco), p. 167.

Suntzeff, N., Kinman, T.D., and Kraft, R.P., 1991, *Astrophys. J.*, **367**, 528.

Suntzeff, N.B. and Kraft, R.P., 1996, *Astron. J.*, **111**, 1913.

Suntzeff, N.B., Schommer, R.A., Olszewski, E.W., and Walker, A.R., 1992, *Astron. J.*, **104**, 1743.

Surdin, V.G., 1979, *Sov. Astron.*, **23**, 648.

Surdin, V.G. and Charikov, A.V., 1977, *Sov. Astron.*, **21**, 12.

Sweigart, A.V. and Gross, P.G., 1978, *Astrophys. J. Suppl.*, **36**, 405.

Sweigart, A.V. and Mengel, J.G., 1979, *Astrophys. J.*, **229**, 624.

Taillet, R., Salati, P., Longaretti, P.-Y., 1996, *Astrophys. J.*, **461**, 104.

Tayler, R.J., 1986, *Q. Jl. R. astr. Soc.*, **27**, 367.

Toomre, A. and Toomre, J., 1972, *Astrophys. J.*, **178**, 623.

Toomre, A., 1977, in *The Evolution of Galaxies and Stellar Populations*, eds. B. Tinsley and R. Larson (Yale University Observatory, New Haven), p. 401

Toth, G. and Ostriker, J.P., 1992, *Astrophys. J.*, **389**, 5.

Tremaine, S., 1981, in *Structure and Evolution of Normal Galaxies*, eds. S.M. Fall and D. Lynden-Bell (Cambridge University Press), p. 67.

Trimble, V. and Leonard, P.J.T., 1996, *Publ. Astron. Soc. Pac.*, **108**, 8.

Tripicco, M.J., 1989, *Astron. J.*, **97**, 735.

Tripicco, M.J., 1993, in *The Globular Cluster–Galaxy Connection*, eds. G.H. Smith and J.P. Brodie (ASP, San Francisco), p. 432.

van Altena, W.F., Lee, J.T., and Hoffleit. E.D., 1991, *The General Catalog of Trigonometric Parallaxes* (Yale University).

VandenBerg, D.A., 1992, *Astrophys. J.*, **391**, 685.

VandenBerg, D.A., Bolte, M., and Stetson, P.B., 1990, *Astron. J.*, **100**, 445.

VandenBerg, D.A., Bolte, M., and Stetson, P.B., 1996, *Ann. Rev. Astron. Astrophys.*, in press.

van den Bergh, S., 1969, *Astrophys. J. Suppl.*, **19**, 145.

van den Bergh, S., 1975, *Ann. Rev. Astron. Astrophys.*, **13**, 180.

van den Bergh, S., 1990a, in *The Dynamics and Interactions of Galaxies*, ed. Roland Wielen (Springer, Berlin), p. 492

van den Bergh, S., 1990b, *Jl. R. astr. Soc. Can.*, **84**, 60.

van den Bergh, S., 1991a, *Astrophys. J.*, **369**, 1.

van den Bergh, S., 1991b, *Publ. Astron. Soc. Pac.*, **103**, 1053.

van den Bergh, S., 1993a, *Astron. J.*, **105**, 971.

van den Bergh, S., 1993b, in *The Globular Cluster–Galaxy Connection*, eds. G.H. Smith and J.P. Brodie (ASP, San Francisco), p. 346.

van den Bergh, S., 1993c, *Astrophys. J.*, **411**, 178.

van den Bergh, S., 1994, *Astron. J.*, **108**, 2378.

van den Bergh, S., 1995a, *Astrophys. J.*, **446**, 39.

van den Bergh, S., 1995b, *Astrophys. J. Lett.*, **451**, L65.

van den Bergh, S., 1995c, *Nature*, **374**, 215.

van den Bergh, S. and Lafontaine, A., 1984, *Astron. J.*, **89**, 1822.

van den Bergh, S. and Morbey, C.L., 1984, *Astrophys. J.*, **283**, 598.

van den Bergh, S., Morbey, C.L., and Pazder, J., 1991, *Astrophys. J.*, **375**, 594.

van den Bergh, S. and Morris, S., 1993, *Astron. J.*, **106**, 1853.

van den Bergh, S., Pritchet, C., Grillmair, C. 1985, *Astron. J.*, **90**, 595.

Vietri, M. and Pesce, E., 1995, *Astrophys. J.*, **442**, 618.

Walker, A.R., 1990, *Astron. J.*, **100**, 1532.

Walker, A.R., 1992a, *Astron. J.*, **103**, 1166.

Walker, A.R., 1992b, *Astrophys. J. Lett.*, **390**, L81.

Walker, I.R., Mihos, J.C., and Hernquist, L., 1996, *Astrophys. J.*, **460**, 121.

Wannier, P. and Wrixon, G.T., 1972, *Astrophys. J. Lett.*, **173**, L119.

Watson, A.W., Gallagher, J.S., Holtzman, J.A., Hester, J., Mould, J.R., Ballester, G.E., *et al.* 1996, *Astron. J.*, **112**, 534.

Webbink, R.F., 1979, *Astrophys. J.*, **227**, 178.

Weinberg, M.D., 1994a, *Astron. J.*, **108**, 1398.

Weinberg, M.D., 1994b, *Astron. J.*, **108**, 1403.

Weinberg, M.D., 1994c, *Astron. J.*, **108**, 1414.

West, M.J., 1993, *Mon. Not. R. astr. Soc.*, **265**, 755.

West, M.J., Côté, P., Jones, C., Forman, W., and Marzke, R.O., 1995, *Astrophys. J. Lett.*, **453**, L77.

Wheeler, J.C., 1979, *Astrophys. J.*, **234**, 569.

Wheeler, J.C., Sneden, C., and Truran, J.W., 1989, *Ann. Rev. Astron. Astrophys.*, **27**, 279.

White, S.D.M. and Rees, M.J., 1978, *Mon. Not. R. astr. Soc.*, **183**, 341.

White, R.E. and Shawl, S.J., 1987, *Astrophys. J.*, **317**, 246.

White, R.E., III, 1987, *Mon. Not. R. astr. Soc.*, **227**, 185.

Whitmore, B.C. 1997, in *The Extragalactic Distance Scale*, eds. M. Donahue and M. Livio (Cambridge University Press), in press.

Whitmore, B.C., Schweizer, F., Leitherer, C., Borne, K., and Robert, C., 1993, *Astron. J.*, **106**, 1354.

Whitmore, B.C. and Schweizer, F., 1995, *Astron. J.*, **109**, 960.

Whitmore, B.C., Sparks, W.B., Lucas, R.A., Macchetto, F.D., and Biretta, J.A., 1995, *Astrophys. J. Lett.*, **454**, L73.

Whitmore, B.C., Miller, B.M., Schweizer, F., and Fall, S.M., 1997, *Astron. J.*, submitted.

Wielen, R., 1985, in *Dynamics of Star Clusters*, eds. J. Goodman and P. Hut (Reidel, Dordrecht), p. 449.

Wielen, R., 1988, in *Globular Cluster Systems in Galaxies*, eds. J.E. Grindlay and A.G.D. Philip (Reidel, Dordrecht). p. 393.

Willson, L.A., Bowen, G.H., and Struck-Marcell, C., 1987, *Comments Astrophys.*, **12**, 17.

Woolley, R.v.d.R., *et al.* 1966, *Roy. Obs. Ann.*, No. 2.

Worthey, G., 1994, *Astrophys. J. Suppl.*, **95**, 107.

Worthey, G., Faber, S.M., and González, J.J., 1992, *Astrophys. J.*, **398**, 69.

Wright, G.S., James, P.A., Joseph, R.D., and McLean, I.S., 1990, *Nature*, **344**, 417.

Yan, L., and Mateo, M., 1994, *Astron. J.*, **108**, 1810.

Zepf, S.E. 1995, in *Dark Matter*, eds. S.S. Holt and C.L. Bennet (AIP, New York), p. 153.

Zepf, S.E., 1996, in *Galaxy Interactions in Pairs, Groups, and Clusters*, ed. G. Longo, in press.

Zepf, S.E. and Ashman, K.M., 1993, *Mon. Not. R. astr. Soc.*, **264**, 611.

Zepf, S.E., Ashman, K.M., and Geisler, D., 1995, *Astrophys. J.*, **443**, 570.

Zepf, S.E., Ashman, K.M., English, J., Freeman, K.C., and Sharples, R.M., 1997, *Astron. J.*, submitted.

Zepf, S.E., Carter, D., Sharples, R.M., and Ashman, K.M., 1995a, *Astrophys. J. Lett.*, **445**, L19.

Zepf, S.E., Geisler, D., and Ashman, K.M., 1994, *Astrophys. J. Lett.*, **435**, L117.

Zepf, S.E. and Silk, J., 1996, *Astrophys. J.*, **466**, 114.

Zinn, R., 1985, *Astrophys. J.*, **293**, 424.

Zinn, R., 1986, in *Stellar Populations*, eds. C.A. Norman, A. Renzini, and M.Tosi (Cambridge University Press), p. 73.

Zinn, R., 1990, *J. R. astr. Soc. Can.*, **84**, 89.

Zinn, R., 1991, in *The Formation and Evolution of Star Clusters*, ed. K. Janes (ASP, San Francisco), p. 532.

Zinn, R., 1993, in *The Globular Cluster–Galaxy Connection*, eds. G.H. Smith and J.P. Brodie (ASP, San Francisco), p. 38.

Zinn, R., 1996, in *Formation of the Galactic Halo...Inside and Out*, eds. H. Morrison and A. Sarajedini (ASP, San Francisco), p. 211.

Zinn, R. and Persson, S.E., 1981, *Astrophys. J.*, **247**, 849.

Zinneker, H., Keable, C.J., Dunlop, J.S., Cannon, R.D., and Griffiths, W.K. 1988, in *Globular Cluster Systems in Galaxies*, eds. J.E. Grindlay and A.G.D. Phillip (Reidel, Dordrecht), p. 603.

Subject index

abundances
  CNO, 10, 50–1, 61, 125–6
  helium, 10, 14, 15, 18
  ratios, 92, 101, 111
  see also metallicity
ages of globular clusters, 6, 9, 14–18, 7, 106, 129, 144
  differences 10, 16–18, 96, 145

bimodality, 38, 60, 96–100, 117–18, 122–3, 147, 148–50
binary stars, 7, 23–5
  fraction of, 7, 25
black holes, 23, 24
blue stragglers, 6
bulge-shocking, 49

cataclysmic variables (CVs), 24
cD galaxies, 77, 87, 129
Cepheid distances, 32, 82–3, 84
chemical homegeneity, 136, 145
closed-box model of chemical evolution, 117–18
color–magnitude diagrams, 5–14, 16, 17, 18, 23, 61, 63, 68
colors of globular clusters, 88–101
  average colors, 88–92, 113, 122–3
  distribution of colors, 94–101, 115–16, 117
  radial gradient, 92–4, 114, 115
concentration parameter, 26, 48, 49, 65
  see also King model
convection, 15
core collapse 24, 26, 29, 41, 48, 49, 62, 65

dark matter,
  galaxies, 101–2
  globular clusters, 29–30, 128
destruction of globular clusters, 124–7, 146, 148
  see also dynamical friction, evaporation
disk-shocking, 41, 42, 44, 45, 48, 49
dwarf elliptical (dE) galaxies, 77, 89, 91
dwarf spheroidal (dSph) galaxies, 77, 89, 91
dynamical evolution, 23, 28, 41, 43, 44–50, 107, 145–6
dynamical friction, 44, 45, 46, 51
dynamical relaxation, 35, 48

efficiency of globular cluster formation, 107, 141, 147
equipartition, 28, 45
evaporation, 37, 44, 45, 48, 49, 125

Fundamental Plane,
  of galaxies, 110
  of globular clusters, 37, 62

galactic winds, 111, 122
galaxy formation models 52, 56, 127–8
  hierarchical, 112–13, 122–3
  isolated collapse, 110–12, 113–14
  see also mergers of galaxies
Giant Molecular Clouds, 48, 132, 134–6, 142
globular cluster luminosity function, 49, 66–7, 74, 78–84, 87, 106–8, 128, 129, 134, 142–3, 148, 150
  of elliptical galaxy globular cluster systems, 78–84, 100, 135
  of the Galactic globular cluster system, 33–4, 35, 38, 134
  of spiral galaxy globular cluster systems, 58–9, 63, 66–7, 70–1, 78–9, 81, 84, 135
globular cluster mass function, 35, 49, 50, 81, 83, 100, 106–8, 129, 134, 142–3

helium burning,
  core, 11
  shell, 11
helium diffusion, 15
horizontal branch, 6, 8–11, 16, 17, 31, 43–4, 53, 54, 62, 64, 68, 145
Hubble Constant, 14
Hubble Space Telescope (HST), 7, 15, 22, 23, 42, 62, 63, 70, 71, 74, 79, 81, 82–4, 89, 99, 100, 103–7, 145, 146, 147, 148
Hubble time, 49, 103, 107, 134
hyrodgen burning,
  core, 5
  shell, 7, 11

instability strip, 6, 8, 11
isochrones (stellar), 15, 16

Jeans mass, 128, 129–30, 133

169

Object index